Stress Testing and Risk Integration in Banks

Stress Testing and Risk Integration in Banks

A Statistical Framework and Practical Software Guide (in Matlab and R)

Tiziano Bellini

ELSEVIER

AMSTERDAM • BOSTON • HEIDELBERG • LONDON
NEW YORK • OXFORD • PARIS • SAN DIEGO
SAN FRANCISCO • SINGAPORE • SYDNEY • TOKYO
Academic Press is an imprint of Elsevier

Academic Press is an imprint of Elsevier
125 London Wall, London EC2Y 5AS, United Kingdom
525 B Street, Suite 1800, San Diego, CA 92101-4495, United States
50 Hampshire Street, 5th Floor, Cambridge, MA 02139, United States
The Boulevard, Langford Lane, Kidlington, Oxford OX5 1GB, United Kingdom

Notices

Knowledge and best practice in this field are constantly changing. As new research and experience broaden our understanding, changes in research methods, professional practices, or medical treatment may become necessary.

Practitioners and researchers must always rely on their own experience and knowledge in evaluating and using any information, methods, compounds, or experiments described herein. In using such information or methods they should be mindful of their own safety and the safety of others, including parties for whom they have a professional responsibility.

To the fullest extent of the law, neither the Publisher nor the authors, contributors, or editors, assume any liability for any injury and/or damage to persons or property as a matter of products liability, negligence or otherwise, or from any use or operation of any methods, products, instructions, or ideas contained in the material herein.

Library of Congress Cataloging-in-Publication Data
A catalog record for this book is available from the Library of Congress

British Library Cataloguing-in-Publication Data
A catalogue record for this book is available from the British Library

ISBN: 978-0-12-803590-0

For information on all Academic Press publications
visit our website at https://www.elsevier.com/

www.elsevier.com • www.bookaid.org

Publisher: Nikki Levy
Acquisition Editor: Scott Bentley
Editorial Project Manager: Susan Ikeda
Production Project Manager: Susan Li
Designer: Mark Rogers

Typeset by SPi Global, India

Dedication

Ai miei genitori, Patrizia e Leonardo

Contents

Tiziano Bellini's Biography

Tiziano Bellini received his PhD degree in statistics from the University of Milan after being a visiting PhD student at the London School of Economics and Political Science. He gained wide risk management experience across Europe, in London, and in New York. He currently holds a senior management position at EY Financial Advisory Services in London. Previously he worked at HSBC's headquarters, Prometeia, and other leading Italian companies. He is a guest lecturer at the London School of Economics and Political Science. Formerly, he served as a lecturer at the University of Bologna and the University of Parma. He has published in the *European Journal of Operational Research*, *Computational Statistics and Data Analysis*, and other top-reviewed journals. He has given numerous training courses, seminars, and conference presentations on statistics, risk management, and quantitative methods.

Preface

This book provides a comprehensive view of stress testing and risk integration in banks. Statistical tools are used as a guide to assess how adverse economic conditions may affect economic and financial resilience. A first focus is on links connecting macroeconomic variables and bank-specific conditions. Time series are studied as part of the scenario analysis that is the very heart of the process. An asset and liability management frame is the other necessary ingredient to represent balance sheet and profit and loss dynamics under stress.

The text is aimed at graduates, master students, and practitioners. Examples and business cases are included in all chapters. The analysis of a stylized bank (i.e., Bank Alpha) highlights the interconnections among the key areas of risk management interest all over the book. As an additional practical aspect, software examples in MALTAB and R enable readers to familiarize themselves with empirical implementations. The choice of these tools is driven by their wide use both in academia and in banks. Their flexibility in dealing with time series, Monte Carlo simulations, and other statistical models is an additional advantage experienced throughout the text.

When dealing with an overall stress testing exercise, one may be dragged into the details of a specific topic. The risk is to lose the broader picture behind it. On this subject, the book presents a global view of a bank. The toolkit encompasses a wide range of instruments such as statistical techniques, accounting rules, and regulatory standards.

A special focus is devoted to bank vulnerabilities. On this subject, it is worth quoting the following from the *Financial Times* (Sep. 14, 2008):

> *Deal or no deal? ... Without a buyer, or collective action by the bank's peers to save it, Lehman seemed poised to become the latest big casualty of this financial crisis. As possible saviors such as Barclays and Bank of America pulled back from talks, the implications for Lehman appeared dire. Bankruptcy would be a tragic ending for an institution known for its fighting spirit.*

Both long- and short-term peculiarities are examined. In this regard, the final goal of the book is to study how to integrate risks and perform a reverse stress test. In line with the need to manage risk interdependencies, top-down and bottom-up risk integration approaches are investigated. More specifically, a bottom-up framework aligned with a wide stress testing mechanism is used to integrate market, credit, interest rate, and liquidity risks through the lenses of an asset and liability management structure. On this, reverse stress testing constitutes a critical instrument to guide a strategic risk management process. Macroeconomic analysis combined with what-if enquiries are cutting-edge tools qualifying a risk management practice operating in a global economy.

The book is structured as follows. Chapter 1 introduces stress testing and risk integration by pointing out their role within a regulatory and managerial framework. As a necessary ingredient of the overall analysis, Chapter 2 introduces the key concepts related to time series analysis. The focus is on statistical techniques such as vector autoregression, vector error-correction, and global vector autoregression models. Forecasting, scenario enrichment, and Monte Carlo simulations are explored as an initial step of the stress testing process. Chapter 3 outlines the key concepts related to asset and liability management. Bank Alpha is described in detail in this chapter. Margin and liquidity risks are studied by consideration of the entire balance sheet structure. Then the focus shifts to the trading book through the proposal of the most common value at risk methods. Chapter 4 illustrates how to model credit portfolio losses. A bridge is built between the portfolio modeling literature and Basel II Accord regulatory capital requirements. Moreover, a macroeconomic transmission mechanism is illustrated as a key process driving balance sheet and economic (profit and loss) projection as detailed in Chapter 5. In this regard, Bank Alpha is extensively used to show how balance sheet projections feed profit and loss, risk-weighted assets, and eventually, capital and liquidity ratios. Chapter 6 collects the results of the preceding chapters to assess capital ratios, leverage, and liquidity indices under stress. Finally, Chapters 7 and 8 describe how to integrate risks and perform a coherent reverse stress testing analysis.

Tiziano Bellini
London, May 2016

Acknowledgments

This book is the fruit of a long journey that started with my graduation in business and economics from the University of Parma, enriched with a doctorate in statistics from the University of Milan and enhanced through experience gained working in the financial industry across Europe, in London, and in New York.

I am grateful to Marco Riani and Luigi Grossi, who first introduced me to statistics and data analysis. I also thank Emeritus Professor Anthony Atkinson, who, through Marco Riani, transmitted to me his love for both statistics and London, where I wrote this book. Special thanks are addressed to Carlo Toffano and Flavio Cocco, who guided my steps through risk management practice. This book benefits from our exchange of opinions to face bank challenges all over Europe.

I thank Glyn Jones and Scott Bentley for supporting and endorsing this project. I am beholden to Susan Ikeda for her continuous help during the publication process. I am also grateful for all comments received from four anonymous reviewers. The book benefits greatly from their challenges and advice.

I am immensely indebted to Katarzyna Marciak, Daniel Ruediger, Charity Muhangi, and Alessandra Luati for their encouragement and precious help in reviewing earlier versions of the manuscript.

My greatest thanks are addressed to my parents, who supported me when I despaired of my ability to combine research and professional activities. Their quiet and continuous help gave me the strength to continue pursuing my ideas even when this appeared to be impossible. I also thank my sister for her warming admiration. I dedicate this work to her and my nephew Leonardo, the most energizing and adorable creature that God donated to my family.

Chapter 1

Introduction to Stress Testing and Risk Integration

Chapter Outline

Since the 2007–09 crisis, increasing attention has been devoted to capital adequacy and balance sheet integrity. Banks have been required to improve the quality of their own funds, strengthen their liquidity structure, and enforce their risk management processes. As a starting point, this chapter outlines the regulatory response to the recent financial crunch. On this subject, stress tests and risk integration are useful tools to enhance bank resilience against adverse conditions. Then, Bank Alpha's illustrative example is introduced to show how an international bank runs its business. It serves to outline throughout the book all complex challenges one needs to face when modeling risks. As in an executive summary, this introductory chapter shows some of Bank Alpha's main stress testing and risk integration results. Finally, a practical guide to explore the text is provided. It serves as a map for the reader looking for orientation during the deep-dive journey.

1.1 ANTIDOTE TO THE CRISIS

A series of failures recently sharpened the question about the role of banks in a modern economic system. On this subject, two ways may be followed to connect savings and investments. Firstly, fund suppliers may directly meet the financial demand by acquiring equity positions or debt instruments. However, the wide range of costs associated with direct finance justifies a second way to link money supply and demand. Financial intermediaries screen, monitor, and diversify risks by providing credit to those needing resources.

It is worth noting that, in the recent past, banks progressively moved from their traditional institutional background to a more marked economic value

Stress Testing and Risk Integration in Banks. http://dx.doi.org/10.1016/B978-0-12-803590-0.00001-1

creation perspective. This evolution raised a possible conflict with their social role by highlighting the potential for systemic breakdown. In this regard, given the nature of their operations, banks never hold sufficient capital to guarantee full deposit withdrawals. Additionally, the opaque nature of financial investments does not allow analysts to distinguish the problems specific to one intermediary from those affecting the industry as a whole. As a result, the distress of one entity may lead to runs on others as well. These are the reasons why laws and regulations govern financial intermediation, as detailed in the following sections.

1.1.1 What Went Wrong

Many economic crises in history originated as failures of financial intermediaries. A few banks became bankrupt during the 2007–09 crisis, and many more had impaired operations. Nevertheless, major disruptions occurred among new segments of financial intermediation. A run on asset-backed commercial paper (ABCP) liabilities was one of the main issues experienced during the recent crisis. These short-term funding instruments were used to finance asset portfolios with long-term maturities. ABCP issuers (conduits) performed a typical financial intermediation function but they were not banks. In many instances, banks were the driving force behind ABCP funding growth. They sponsored these activities and provided the required liquidity. However, this new structure shifted a component of financial intermediation away from its traditional location. Additionally, money market mutual funds experienced a run on their liabilities. This event, in turn, triggered an even bigger run on ABCP issuers. The Lehman Brothers Business Case 1.1 exemplifies some of the above-mentioned financial intermediation failures.

Business Case 1.1   Lehman Brothers
Since the early 1900s Lehman Brothers had developed its banking practice, becoming a well-known investment bank. In the first decade of this century, the company widely expanded its services. The most complex products developed in the wake of the financial deregulation belonged to its business. During the period from 2006 to 2007, Lehman Brothers initiated a new strategy. The company aggressively bought real-estate assets. At the end of its 2007 fiscal year, Lehman Brothers held $111 billion in commercial or residential assets and securities, more than double the $52 billion it held at the end of 2006, and more than four times its equity. Market illiquidity and massive losses in this business caused rating agencies and investors to express concerns in 2008. During that summer, the financial situation of Lehman Brothers becomes unsustainable. The crisis became public. On the weekend of Sep. 12–14, 2008, the government communicated its intention not to bail out the firm. A meeting with the major Wall Street investment banks was organized. A private solution similar to that for the Long-Term Capital Portfolio L.P. structured in 1998 was planned. Despite the interest from Bank of America and Barclays, the discussions ultimately failed. Suddenly, Lehman

Brothers realized that it would not be able to raise enough funds to open for business the next day. Its board of directors voted to file for Chapter 11 bankruptcy protection on Sep. 15, 2008. The bank was granted an opportunity to have certain parts of its operations dismantled in an orderly fashion overseen by a bankruptcy court.

The Lehman Brothers case revealed the major issues encountered during the crisis:

- **Level and quality of capital (capital ratios).** The crisis showed there was an insufficient level of regulatory capital to cover losses and write-downs of some banks. Inconsistencies in the definition of *capital* across countries prevented markets from assessing banks' capital quality. A number of banks continued to make large distributions. There were dividend payments, share buybacks, and generous compensation payments even though financial conditions were deteriorated. These losses destabilized the banking sector and exacerbated a downturn in the real economy.
- **Risk coverage of capital (leverage).** High leverage operations and complex securitization exposures were major sources of losses for many banks. Nonetheless, their capital framework did not substantially capture these risks.
- **Illiquidity (liquidity ratios).** Before the crisis, funding was readily available at low cost. The rapid reversal of market conditions highlighted how quickly liquidity could vanish and this could last for an extended period. Many banks experienced difficulties managing their liquidity. Central banks needed to take action to support both the functioning of money markets and, in some cases, individual institutions. Business Case 1.2 focuses on Northern Rock's causes of distress.

Business Case 1.2   Northern Rock

Northern Rock was a building society (i.e., a mutually owned savings and mortgage bank) until its decision to go public in 1997. This bank was a regionally based institution, serving its local customers. Its success was mainly related to the revitalization of the North East of England following the decline of traditional industries, such as coal mining and shipbuilding. Northern Rock had larger ambitions. In the 9 years after it had gone public to the eve of the crisis in Jun. 2007, its total assets grew from £17.4 billion to £113.5 billion. Its liability structure reflected this unusual dynamic by depending heavily on nonretail funds. By the summer of 2007, only 23% of its liabilities were in the form of retail deposits. The rest of its funding came from short-term borrowing in the capital markets, or through securitized notes. ABCP was the favored means to fund Northern Rock's growth.

In the summer 2007 the subprime crisis started affecting banks' balance sheets. The demise of Northern Rock dated from Aug. 9. The news that BNP

Paribas was closing three off-balance sheet investment vehicles, with exposures to US subprime mortgage assets, caused panic in the market. This occurred after other investment vehicles experienced difficulties in rolling over their short-term borrowing. On Aug. 9, the European Central Bank injected €94 billion into the European banking system. A mix of unusual leverage and poisoned financial instruments caused the Northern Rock liquidity crisis. In Sep. 2007 an old-fashioned bank run occurred. Television viewers around the world witnessed depositors standing in line outside Northern Rock branch offices waiting to withdraw their money.

A regulatory response was imperative. Amendments to ongoing rules were discussed and implemented, as described in the next section.

1.1.2 Regulatory Responses

As a response to the financial crisis, the US regulator introduced the Dodd-Frank Wall Street Reform and Consumer Protection Act 2010 (Dodd-Frank Act, Pub. L. 111-203). A more comprehensive and global solution was proposed by the Basel Committee on Banking Supervision (BCBS) through the so-called Basel III accord (BIS, 2011). The Basel Committee on Banking Supervision does not possess any formal supranational supervisory authority. Nonetheless, all major countries align their regulatory standards to the Basel Accords. In particular, after the so-called Basel I (BIS, 1988), there was a new accord named *Basel II* (BIS, 2006) that introduced a three-pillar regulatory structure based on minimum capital requirements (pillar 1), regulatory supervision risk management (pillar 2), and market disclosure (pillar 3).

Basel II introduced three main innovations. Firstly, a refinement on risk buckets for the capital adequacy calculations. Secondly, internal risk-based models became a new paradigm to assess banks' minimum capital requirements. Thirdly, operational risk was adopted as an additional pillar 1 component. The financial crisis, however, induced the international community to start discussing the creation of new regulatory standards. G20 leaders committed to the development of internationally agreed rules to improve banking regulations.

Basel III did not revolutionize the structure of the Basel II accord. The vast bulk of Basel II remains in force. Indeed, no substantial changes occurred concerning the discipline regarding credit and operational risks. Nonetheless, a reform characterized the trading book. Counterpart risk and credit valuation adjustments were strengthened. Additionally, the following key changes enhanced the overall regulatory architecture:

- **Capital ratios.** The capital of a bank serves as a buffer against unexpected losses. Risk-based capital ratios are used to highlight the relationship between capital and risk-weighted assets (RWAs) as detailed below:

$$\text{Capital ratio} = \frac{\text{Regulatory capital}}{\text{RWAs}}. \tag{1.1}$$

A few different definitions of capital (i.e., numerator in Eq. 1.1) may be used for regulatory purposes. Basel III relies on three layers: common equity (core tier 1 capital), additional tier 1 capital, and tier 2 capital. Common shares and retained earnings are the most important own fund items entering into the regulatory capital definition. Innovative and noninnovative instruments belong to a broader tier 1 capital profile. Subordinated debts and similar instruments belong to a more diluted tier 2 capital layer. In contrast, the idea behind RWAs (i.e., the denominator in Eq. (1.1)) is to assign greater weights to riskier exposures (e.g., a credit toward a high-standing central government should have a lower weight compared with a high-risk newly formed private firm). In line with pillar 1, the denominator in Eq. (1.1) consists of the sum of market, credit, and operational RWAs. Chapters 3–5 will consider their computational details. Table 1.1 summarizes the Basel III road map. It anticipates some of the key concepts that will be examined throughout this book. In particular, some hints are provided on the regulatory capital definition and the calculation of risk-based ratios discussed in Chapter 6.

TABLE 1.1 Basel Committee on Banking Supervision Road Map for Minimum Capital Requirements (The Dates Refer to Jan. 1)

	2016	2017	2018	2019
Minimum common equity capital ratio (%)	4.50	4.50	4.50	4.50
Capital conservation buffer (%)	0.625	1.25	1.875	2.50
Minimum common equity plus capital conservation buffer (%)	5.125	5.75	6.375	7.00
Minimum tier 1 capital (%)	6.00	6.00	6.00	6.00
Minimum total capital (%)	8.00	8.00	8.00	8.00
Minimum total capital plus conservation buffer (%)	8.625	9.25	9.875	10.50

- **Leverage ratio.** The leverage ratio was introduced in Basel III as an additional requirement. This indicator pursues the goal of preventing banks from expanding their assets without limit. It is defined as follows:

$$\text{Leverage ratio} = \frac{\text{Capital exposure}}{\text{Total exposure}}, \tag{1.2}$$

where the capital exposure is aligned with the above tier 1 capital exposure, while the total exposure is derived starting from the (financial reporting) asset balance sheet. Adjustments are applied to include off-balance items and exclude some credit risk mitigations. All in all, during the parallel run period (i.e., from 2013 to 2017) a minimum ratio of 3% is tested.

- **Liquidity requirements.** Basel III also introduced constraints as a response to the liquidity issues experienced during the 2007–09 crisis (e.g., Northern Rock). Two complementary perspectives are considered: the liquidity coverage ratio (LCR) aims to check the effectiveness of a liquid asset buffer for short-term liquidity coverage; the net stable funding ratio (NSFR) focuses on the balance between cash outflows and inflows over a longer period. In this regard, let us start from the LCR as given below:

$$\text{LCR} = \frac{\text{High-quality liquid assets}}{\text{Total net cash outflows}}, \tag{1.3}$$

where a prescriptive list of weights is applied to assets and outflows. As an example, central bank cash reserves are assigned a weight of 100% as part of the numerator. In contrast, cumulative outflows over a 30-day stress period are the basis of the denominator in Eq. (1.3).

The NSFR aims at inducing banks to balance maturity mismatches as follows:

$$\text{NSFR} = \frac{\text{Available stable funding}}{\text{Total net cash outflows}}, \tag{1.4}$$

where stable funding is the portion of equity and liabilities expected to provide reliable sources over a 1-year time horizon. The required stable funding is measured by the taking into account of the characteristics of liquidity of assets and off-balance sheet exposures. All weights used in this computation are listed as regulatory requirements.

As an additional component of the overall process to strengthen banking resilience, regulators rely heavily on stress testing. This process has become a common tool, all over the world, to assess a bank's capability to face its obligations under adverse economic scenarios. In the United States, the Comprehensive Capital Analysis and Review was introduced as a key component of the Dodd-Frank Act and is an annual exercise conducted by the Federal Reserve on the largest bank holding companies operating in the United States (FRB, 2015). In Europe, the European Central Bank (together with the European Banking Authority) deployed a first wide stress testing exercise in 2011 that was then replicated in 2014 and 2016 (EBA, 2016). The Bank of England in 2014 aligned with the Federal Reserve Board in committing to an annual stress testing process for major UK banks (BOE, 2015). Other regulators across the world followed the same approach by introducing a stress testing assessment on a regular basis.

Bank Alpha is introduced in the next section as an example of a stylized commercial bank operating on a multicountry basis. It highlights the key tools and outputs one aims to investigate during a stress testing and risk integration process.

1.2 STRESS TESTING, RISK INTEGRATION, AND REVERSE STRESS TESTING

A stress test is a comprehensive process where statistical tools are used as a guide to assess how adverse macroeconomic scenarios may affect a bank resilience. In this regard, a standardized and highly structured uniform process used by regulators to compare firms operating in the financial industry may fail to capture some of the specificities that emerged during the recent financial crisis. A more detail-oriented way to assess risks should be followed. Risk integration and reverse stress testing should be used to highlight bank peculiarities both in the long term and in the short term. As anticipation of the next chapters' results, Bank Alpha is explored in the following sections to uncover some of the key details about the stress testing exercise as well as the risk integration and reverse stress testing processes.

1.2.1 Stress Testing

Bank Alpha is an international bank operating in the following main areas: United States, China and developing Asia, the euro area, Japan, and the United Kingdom. Table 1.2 summarizes its balance sheet at t_0 (i.e., stress testing starting point).

A $70 billion (loan) commercial portfolio qualifies Bank Alpha's business. Securities account for $14 billion and cash resources for $8 billion. Other activities represent the residual $8 billion out of total assets of $100 billion. On the liability side, deposits of $70 billion highlight Bank Alpha's capability to borrow from its customers. Additionally, the bank relies on other liabilities,

TABLE 1.2 Bank Alpha's Balance Sheet at t_0 ($ Billions)

Assets	Value	Liabilities	Value
Cash resources	8.00	Deposits	70.00
Securities	14.00	Other liabilities	17.00
Loans	70.00	Subordinated debts	4.00
Other assets	8.00	Noncontrolling interests	2.00
		Shareholder equity	7.00
Total assets	100.00	Total liabilities	100.00

subordinated debts, and noncontrolling interests. Finally, shareholder equity stands for the bank core own resources.

With regard to equity, a distinction is made between accounting and regulatory perspectives. As anticipated earlier, capital ratios are computed as the proportion between regulatory capital and risk-weighted assets. Section 1.1.2 showed that the regulatory capital is spread across three main layers: common equity tier 1 (CET1), additional tier 1 capital, and tier 2 capital. The key difference among these components is rooted in the degree of capital dilution. Common shares and retained earnings are the key components of the common equity. Additional tier 1 and tier 2 instruments belong to a less forceful own fund definition. Table 1.3 summarizes Bank Alpha's regulatory capital at t_0.

The capital ratio's denominator (i.e., RWAs) is the sum of market, credit, and operational RWAs. Starting from the $7 billion shareholder equity (common shares $4 billion, preferred shares $1 billion, retained earnings $2 billion),

TABLE 1.3 Bank Alpha's Regulatory Capital at t_0 ($ Billions)

Regulatory Capital	Value
CET1	4.44
Additional tier 1	1.00
Tier 1	5.44 (subtotal)
Tier 2	6.00
Total capital	11.44 (total)

CET1, common equity tier 1.

TABLE 1.4 Bank Alpha's Risk-Weighted Assets at t_0 ($ Billions)

RWAs	Value
Market risk	2.31
Operational risk	8.93
Credit risk	54.80
Other risks	1.20
Shortfall	2.50
Add-on RWAs	6.26
Total RWAs	76.00

RWAs, Risk-weighted assets.

regulatory adjustments listed in Chapter 6 are applied. Table 1.4 summarizes Bank Alpha's RWAs at t_0.

The comparison of common equity and RWAs highlights a common equity ratio of 5.84%. In contrast, the tier 1 ratio is 7.16% and total capital ratio is 15.05%. At t_0, Bank Alpha exceeds the minimum 4.50% common equity ratio, the 6.00% tier 1 ratio and the 8% total capital ratio detailed in Table 1.1.

It is worth mentioning that additional buffers are required for stress testing purposes. Fig. 1.1 points out that the tier 1 and the total capital ratios are the very bottom threshold a bank needs to align with. Then, specific buffers address pillar 2, systemic, and stress test-specific requirements.

From a structural point of view, a bank should align with the minimum 3% leverage ratio (i.e., ratio between capital exposure and total asset exposure). Table 1.5 shows that Bank Alpha exceeds the minimum threshold at t_0.

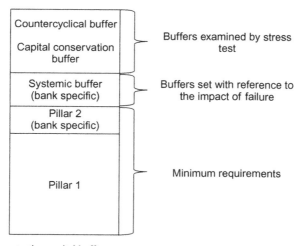

FIG. 1.1 Stress testing capital buffers.

TABLE 1.5 Bank Alpha's Leverage Ratio at t_0 ($ Billions)

	Value
On-balance sheet items	100.00
Assets deducted from tier 1 capital	−1.56
Total balance sheet exposures	98.44
Tier 1 capital	5.44
Leverage ratio	5.53%

Additionally, a bank needs to hit the minimum liquidity standards in terms of the LCR and the NSFR. Bank Alpha's LCR at t_0 is 104.14% and its NSFR is 88.25% (see details in Chapter 6).

Once Bank Alpha's structure at t_0 has been detailed, the next step is to analyze the economic evolution under stress. A scenario represents a coherent macroeconomic path designed to assess bank resilience. An illustrative 3-year stress path is used (throughout the book) to assess Bank Alpha's capability to face its obligations. The focus of the analysis is on Bank Alpha's main business areas (the United States, China and developing Asia, the euro area, Japan, and the United Kingdom).

As summarized in Fig. 1.2, this scenario relies on a generalized contraction of the major economies. It takes the form of a steep reduction in GDP growth in all major Western countries and a fall in the historically high Chinese growth rate. More specifically, the top-left panel in Fig. 1.2 highlights very low US interest rates. This applies both to the short-term rates (i.e., US r^{ST}=0.1%) and to the long-term rates (i.e., US r^{LT} increases from 1% to 2% during the 3-year scenario). This path is accompanied by a moderate increase in US unemployment. The top-right panel in Fig. 1.2 shows the high correlation between US GDP growth and GDP growth in the other economies. The shape and growth levels are substantially aligned (with the exception of the Chinese

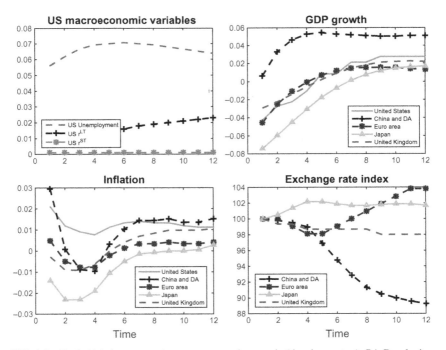

FIG. 1.2 Bank Alpha's stress testing macroeconomic scenario (time in quarters). *DA*, Developing Asia.

economy) throughout the whole period. Furthermore, the bottom-left panel describes how inflation behaves during the stress test. Finally, the bottom-right panel shows the movements in exchange rates with regard to the US dollar.

The need to bridge the macroeconomic evolution and bank risk indicators is imperative. A transmission mechanism is then used to assess stressed capital ratios. As an introductory summary, Table 1.6 shows how Bank Alpha is affected by the adverse scenario described above. The common equity ratio breaches the minimum 4.50% Basel III threshold at t_1 and t_2, while the tier 1 ratio stays below the 6.00% threshold at t_1 and t_2. These results would cause an immediate regulatory intervention followed by a series of redemption actions such as the raising of capital and business restructuring.

Table 1.7 outlines Bank Alpha's solidity in terms of the leverage and liquidity ratios.

TABLE 1.6 Bank Alpha's Capital Ratios During the 3-year Stress Testing Exercise ($ Billions)

		Stress		
	t_0	t_1	t_2	t_3
CET1	4.44	3.57	3.79	4.44
Tier 1	5.44	4.57	4.79	5.44
RWAs	76.00	108.00	95.81	88.03
CET1 ratio	5.84%	3.31%	3.96%	5.04%
Tier 1 ratio	7.16%	4.23%	5.00%	6.18%

CET1, common equity tier 1; RWAs, Risk-weighted assets.

TABLE 1.7 Bank Alpha's Leverage and Liquidity Ratios During the 3-year Stress Testing Exercise

		Stress		
	t_0	t_1	t_2	t_3
Leverage ratio (%)	5.53	4.41	4.44	4.84
LCR (%)	104.14	100.60	102.97	104.13
NSFR (%)	88.25	93.27	92.15	90.46

LCR, Liquidity coverage ratio; NSFR, net stable funding ratio.

A major drawback of the current pillar 1 regulatory framework is not to explicitly rely on integrated risk measures. The 2007–09 crisis highlighted the importance of risk interdependencies. Thus, aiming to fill this gap, Chapters 7 and 8 will describe a risk integration framework to support the risk management activity for both internal and regulatory purposes (e.g., reverse stress testing, internal capital adequacy process). The next section provides some hints on this topic by focusing on Bank Alpha's key results.

1.2.2 Risk Integration and Reverse Stress Testing

Despite the awareness of interconnections among risk sources, the current regulatory framework still focuses on a silos approach to assess capital requirements. Pillar 1 market, credit, and operational risks are computed independently. Then they are summed to obtain the overall RWAs. Other risks (e.g., pillar 2) are taken into account to assess additional capital buffers.

The regulatory persistence in avoiding integrated risk measures for the purpose of pillar 1 did not prevent techniques being developed. Two main methods have been implemented in the last few years: top-down and bottom-up modeling:

- **Top-down modeling.** The idea behind a top-down approach is to start from the marginal distribution of individual risks and aggregate them through a joint distribution function. Some authors use elliptical distributions, while others use copulas. The difficulties in implementing a top-down solution are twofold. On the one hand, data scarcity makes the choice of a joint distribution function, to some extent, arbitrary. On the other hand, model parameters are usually estimated on the basis of very few observations. Starting from the studies conducted by Kuritzkes et al. (2002), Dimakos and Aas (2004), and Aas et al. (2007) one may realize that the representation of risk interaction is problematic when starting from risk silos. For these reasons, researchers as well as practitioners turned their attention to a bottom-up approach. This angle is very close to the stress regulatory framework outlined in the previous section. Therefore the stress testing analysis may be considered as a risk integration forerunner, as shown below.
- **Bottom-up modeling.** The goal pursued by a bottom-up approach is to tackle micro-level interactions linking macroeconomic variables, risk factors, and individual financial instruments. In the recent literature, Grundke (2009) applied sampling techniques to integrate market and credit risks by using the well-known CreditMetrics solution. This framework was additionally used to evaluate the accuracy of top-down approaches (Grundke, 2010). Kretzschmar et al. (2010) implemented an analysis based on the balance sheet of a composite European bank. Their study highlighted that modular approaches can lead to a bank being undercapitalized. From a different perspective, Drehmann et al. (2010) proposed a framework for a holistic stress testing. Banking microstructures were investigated from an

asset and liability point of view. The comprehensive impact of credit and interest rate risks were assessed in terms of a bank's economic value and capital adequacy. From the same perspective, Alessandri and Drehmann (2010) focused on earnings at risk model. They showed that interactions matter when one is assessing banking economic capital. Bellini (2013) highlighted the effectiveness of all the above-mentioned approaches in capturing some specific integration issues. Chapters 7 proposes a framework where each of these models might fit in. This system highlights the interaction between long- and short-term banking solvency. Its main strength is to investigate economic capital and liquidity with use of the same modeling platform.

Chapters 7 and 8 will consider the risk integration details. For the purpose of this introduction, it is worth summarizing the key outputs of a process based on a wide bottom-up approach where long-term (i.e., economic capital) and short-term (i.e., liquidity mismatching) solvency are explored contemporaneously. In both cases a loss function is specified. The long-term solvency analysis focuses on market, interest rate and credit risk joint estimation. Starting from Monte Carlo simulations, a loss function captures interdependencies based on a micro-level modeling. Liquidity mismatches are also explored in each scenario to assess the bank's capability to face adverse circumstances. In both cases a value at risk or expected shortfall is used as a synthetic measure based on a comprehensive loss function.

As an example, Bank Alpha is used to describe how the integration process works. Table 1.8 summarizes the economic capital (long-term solvency) analysis. The assets and debits columns in Table 1.8 highlight that both balance sheet sides contribute to the overall loss. Additionally, the marginal contribution of default is incomparably higher than that of the interest rate.

TABLE 1.8 Bank Alpha's Integrated Economic Capital ($ Billions)

		Assets		
	Total	Default	Interest	Debits
$EC_{VaR,99.9\%}$	8.65	8.50		0.15
		8.95	−0.45	
$EC_{ES,99.9\%}$	8.90	8.70		0.20
		9.30	−0.60	

Economic capital value at risk at the 99.9% confidence level ($EC_{VaR,99.9\%}$) and economic capital expected shortfall at the 99.9% confidence level ($EC_{ES,99.9\%}$).

TABLE 1.9 Bank Alpha's Liquidity Mismatch ($ Billions)

	Total
$LM_{VaR,99.9\%}$	−8.90
$LM_{ES,99.9\%}$	−9.15

Liquidity mismatch value at risk at the 99.9% confidence level ($LM_{VaR,99.9\%}$) and liquidity mismatch expected shortfall at the 99.9% confidence level ($LM_{ES,99.9\%}$).

The overall economic capital value at risk at the 99.9% confidence level of $8.65 billion shows a potential $108.12 billion of RWAs[1] ($111.25 billion in the case of the economic capital expected shortfall at the 99.9% confidence level). Therefore Table 1.8 suggests that a holistic measure may be even more conservative than the worst stress testing outcome. Indeed, Table 1.6 showed a maximum $108.00 billion RWAs during the 3-year stress testing exercise.

Additionally, the comprehensive liquidity analysis highlights that Bank Alpha is subject to critical issues. More specifically, the liquidity mismatch (Table 1.9) value at risk at the 99.9% confidence level of −$8.90 billion (−$9.15 billion liquidity mismatch expected shortfall at the 99.9% confidence level) represents the excess of cash outflows over cash inflows during a short period (i.e., 3 months). Stated differently, if the liquidity analysis summarized in Table 1.7 suggested no issues in terms of liquidity (i.e., the LCR and NSFR were reasonably high), the integrated analysis pinpoints some potential weaknesses due to deposit instability. Current accounts and other deposits without maturity may be less stable than expected during a crisis, which may seriously threaten the bank.

As a last step of the stress testing and risk integration journey, a reverse stress testing process pursues the goal of finding circumstances that could cause a bank business to fail. Firstly, a reverse stress testing objective function needs to be defined. Then, a framework detailing how to conduct the reverse stress testing is required. In this regard, two main areas of uncertainty are analyzed: macroeconomic scenarios and internal circumstances under which insolvency may occur. The objective function may be defined on the basis of the bank's capability to face its obligations both in the long term and in the short term. The analysis focuses on two main aspects. On the one hand, one explores a macroeconomic surface by considering the most threatening scenarios. Then,

---

1. A connection is drawn between integrated economic capital and the regulatory capital requirement. The RWAs are assumed to be 12.5 times the corresponding economic capital.

one examines a what-if qualitative path by describing the most important internal risk sources (e.g., name concentration and sector clustering).

Bank Alpha's reverse stress testing reveals issues additional to those detected via regulatory stress test and bottom-up risk integration. Name and sector concentration in the credit portfolio accompany threats in the trading book where opaque instruments without a liquid market should cause severe losses to the bank.

All the above results give the flavor of the complexities one needs to bear in mind when dealing with stress testing and risk integration. The reader is invited to move to the next section to understand the key choices made by the author to represent a comprehensive process and structure a practical software guide.

1.3 BOOK STRUCTURE AT A GLANCE

The previous section anticipated some of the main goals of this book by highlighting the connections between the internal processes of a bank and its economic surroundings. A functional representation is given in Fig. 1.3. This structure combined with the modeling workflow shown in Fig. 1.4 provides a comprehensive view of the interdependencies to be analyzed throughout the book.

The left-hand side of Fig. 1.3 represents the transmission channel connecting the overall economy and a specific bank asset and liability management structure. A distinction is made among the trading book, banking book, liabilities, and own funds. Financial instruments actively traded in the market usually belong to the trading book. Fig. 1.3 highlights the market RWAs as a primary synthetic risk measure for this portfolio. The banking book and credit RWAs are aligned in the same column of Fig. 1.3. The banking book encompasses activities that are not actively traded and are meant to be held until their maturity. In a commercial bank, attention is mainly paid to the credit life cycle and its corresponding risk measure: the RWAs. Operational losses characterizing the entire asset and liability management feed the operational risk RWAs. Moreover, assets and liabilities are jointly scrutinized to assess balance sheet dynamic, profit and loss evolution, and liquidity structure. Own funds movements are studied as a critical ingredient to compute capital ratios at inception and during a stress testing exercise. Finally, the right-hand side of Fig. 1.3 brings together all items required to compute capital and liquidity ratios. All in all, the trading

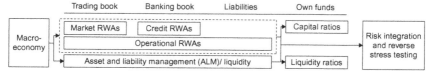

FIG. 1.3   Book structure from a bank functional perspective. *RWAs*, Risk-weighted assets.

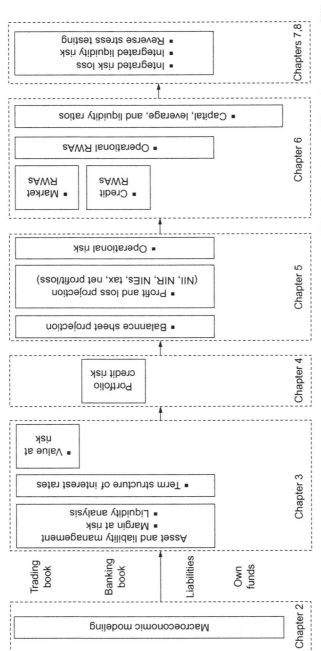

FIG. 1.4 Book plan: intersection of modeling and functional perspectives. *NIEs*, Noninterest expenses; *NII*, net interest income; *NIR*, net interest revenue; *RWAs*, risk-weighted assets.

book, banking book, liabilities, and own funds are the key ingredients for stress testing, risk integration, and reverse stress testing.

The above workflow may be analyzed in conjunction with Fig. 1.4, which maps each chapter's content against the stress testing and risk integration areas listed in Fig. 1.3. The road map begins with Chapter 2, where macroeconomic modeling techniques are investigated. This is a crucial point for the entire book because the concepts and analysis introduced in this chapter are used throughout the entire journey. Chapter 3 outlines the key concepts related to the asset and liability management. Bank Alpha is firstly investigated in detail throughout this chapter. Margin at risk and liquidity analysis rely on the overall balance sheet. It encompasses all assets and liabilities. A more detail discussion focuses on the trading book. The basic ideas behind value at risk methods are described. Stepping forward, Chapter 4 illustrates how to model credit portfolio losses. A bridge is built between the portfolio modeling literature and Basel II Accord regulatory capital requirements. Moreover, a macroeconomic transmission mechanism is illustrated as a key process driving balance sheet and economic (profit and loss) projection as detailed in Chapter 5. In this regard the central part of Fig. 1.4 shows that balance sheet projections feed profit and loss, RWAs, and eventually, capital and liquidity ratios. Bank Alpha is extensively used to show these dynamics. More specifically, Chapter 6 collects as inputs the results obtained in the previous chapters, ending with regulatory assessment. Capital ratios as well as leverage and liquidity indices are extensively examined. The last two chapters describe how to integrate risks and perform a coherent reverse stress testing exercise.

In the next section, each chapter is scrutinized by focusing on the functional relationship with the key banking areas described above (i.e., the trading book, banking book, liabilities, and own funds).

1.3.1 Organization of the Book

The high-level picture described in the previous section serves as a guide to describe the detailed structure of this book. In what follows, a matrix representation is used to couple the banking areas (i.e., the trading book, banking book, liabilities, and own funds) with the risk models studied in each chapter.

Practical software examples and solved exercises are one of the key distinctive features of this text. The choice of MATLAB and R is rooted in their wide use among practitioners as well as their interchangeability and complementarity.

Hints on robust statistics are provided in the appendix in Chapters 2–4. Specific attention is devoted to the forward search of Atkinson et al. (2004). Unlike other robust techniques, the forward search aims to detect aberrant observations with suggestions for model enhancement by use of a graphical approach (Atkinson and Riani, 2000). The key advantage of the forward search is that it is easy to interpret and implement. It does not require any additional assumptions apart from those that characterize the model under analysis.

TABLE 1.10 Main Contents of Chapter 2: Models Used to Fit and Project Macroeconomic Variables

Area	Model	
	AR, MA, ARMA	**VAR, VEC, GVAR**
Macroeconomic modeling	Mathematical description and worked examples in MATLAB and R Box-Jenkins analysis	
		Simulation
		Forecast
		Conditional forecast

AR, Autoregression; ARMA, autoregression moving average; GVAR, global vector autoregression; MA, moving average; VAR, vector autoregression; VEC, vector error-correction.

- **Chapter 2.** External conditions are a key driver for stress testing, risk integration, and reverse stress testing exercises. Chapter 2 introduces a few statistical tools used to model macroeconomic time series. As a starting point, autoregression and moving-average processes are studied from a univariate perspective. Attention is then turned to multivariate series. Vector autoregression and vector error-correction models (Johansen, 1996) are scrutinized. The final part of the chapter focuses on the global vector autoregression model (Pesaran et al., 2004). Table 1.10 summarizes the key topics covered. Worked examples in MATLAB and R accompany an essential mathematical description of all the above models. Box-Jenkins analysis is applied to both univariate and multivariate time series. Moreover, simulation, forecast, and conditional forecast (i.e., forecast conditional on regulatory stress testing scenario) are explored in the multivariate macroeconomic modeling space (i.e., vector autoregression, vector-error correction, and global vector autoregression models).
- **Chapter 3.** An asset and liability management discussion dominates Chapter 3. Bank Alpha is used as illustrative example to familiarize readers with the basic concepts behind a balance sheet representation and its treatment from an asset and liability management standpoint. Table 1.11 highlights the three main modeling areas covered throughout this chapter: margin at risk, value at risk, and liquidity analysis. Apart from the value at risk, which focuses on the trading book, all other techniques apply to the entire asset and liability spectrum.
- **Chapter 4.** This chapter explores the banking book. More precisely, given the nature of a commercial bank, the focus is on credit risk. A portfolio perspective is followed and regulatory capital approaches are investigated to compute credit RWAs. As detailed in Table 1.12, this chapter studies the link connecting macroeconomic variables and credit risk parameters. All in all,

TABLE 1.11 Main Contents of Chapter 3: Models Used for Asset and Liability Management, Value at Risk

Area	Model		
	Margin at Risk	**Value at Risk**	**Liquidity Analysis**
Trading book	Methodological description, guided examples, term structure analysis, and application to Bank Alpha	Methodological description, software-guided examples, application to Bank Alpha	Methodological description, guided examples, application to Bank Alpha
Banking book			
Liabilities			

TABLE 1.12 Main Contents of Chapter 4: Models Used for Credit Risk Modeling

Area	Model		
	Credit Portfolio Modeling	**Credit Risk Parameter**	**Credit RWAs**
Trading book			
Banking book	Methodological description, software-guided examples, application to Bank Alpha		
Liabilities			

RWAs, Risk-weighted assets.

a methodological description is coupled with implementation examples as well as Bank Alpha analysis.

- **Chapter 5.** Balance sheet projection and profit and loss evolution are the key topics analyzed in this chapter. The credit life cycle is outlined as one of the major topics that regulators are keen to monitor under a stressed scenario. The interaction connecting performing and nonperforming portfolios with collective and specific provisions is examined as a crucial component of the overall stress testing architecture. Then the entire balance sheet evolution is explored. Simple rules are shown to project the banking book, trading book, and liabilities. Moving to the profit and loss statement, our attention is first of all concentrated on net interest income as a balance of interest revenue and

TABLE 1.13 Main Contents of Chapter 5: Models Used for Balance Sheet, Profit and Loss Projection and Operational Risk

Area	Model		
	Balance Sheet Evolution	Profit and Loss Evolution	Operational Risk
Trading book	Methodological description, software-guided examples, application to Bank Alpha over a 3-year period (i.e., t_1, t_2, t_3)		
Banking book			
Liabilities			

expenses over a given time horizon (e.g., 1 year). Noninterest revenue due to commission, fees, and so on is examined in combination with noninterest expenses. The aggregation of net interest income, noninterest revenue, and noninterest expenses leads to the preprovisioning net revenue. Once again, credits enter the scene. Loan impairment charges are subtracted from the preprovisioning net revenue to estimate the net profit (loss). Table 1.13 highlights the use of Bank Alpha to show how to combine the balance sheet with profit and loss projections.

- **Chapter 6.** This chapter analyzes own funds and capital ratios from a regulatory standard standpoint (with the main focus on Basel III innovations). All stress testing inputs feed the RWA framework. Market, credit, and operational risks are aggregated in a stress testing framework. Leverage and liquidity ratios are also investigated during a stress testing exercise. Table 1.14 highlights the central role of Bank Alpha to show the implementation process.

- **Chapter 7.** There has been growing interest in risk integration from regulators and practitioners in the last few years. As detailed in Table 1.15, Chapter 7 presents an integrated general framework through which one can assess bank solvency both in the long term (economic capital) and in the short term (liquidity mismatching). The trading book, banking book, and liabilities are part of the architecture. Bank Alpha is used as an example to show process practicalities and main outcomes.

- **Chapter 8.** For a bank exposed to multiple risk factors, many different combinations of stress might result in similar losses. The willingness of a bank to face its obligations may be affected by its portfolio composition, its asset and liability structure, and external macroeconomic conditions, as detailed in Table 1.16. The definition of a reverse stress testing function is one of the first challenges faced. Then much of the arbitrariness of the unstructured reverse stress testing approaches is removed through the proposal of a framework based on the interaction of bank-specific features and external macroeconomic conditions causing a bank to collapse.

TABLE 1.14 Main Contents of Chapter 6: Models Used for Capital and Liquidity Ratios

Area	Model			
	Own Funds Computation	RWA Silos Computation	RWA Aggregation and Capital Ratios	Leverage and Liquidity Ratios
Trading book		Market RWAs under stress (i.e., t_1, t_2, t_3)	Aggregated RWAs under stress (i.e., t_1, t_2, t_3)	Leverage and liquidity ratios under stress (i.e., t_1, t_2, t_3)
Banking book		Credit RWAs under stress (i.e., t_1, t_2, t_3) — Operational RWAs under stress (i.e., t_1, t_2, t_3)		
Liabilities				
Own funds	Regulatory definition			

RWA, Risk-weighted asset.

TABLE 1.15 Main Contents of Chapter 7: Models Used for Risk Integration

Area	Model		
	Bank Solvency Definition	**Risk Integration Modeling**	**Statistical Model Implementation**
Trading book	Bottom-up methodological description, software-guided examples, application to Bank Alpha		
Banking book			
Liabilities			

TABLE 1.16 Main Contents of Chapter 8: Models Used for Reverse Stress Testing

Area	Model		
	Reverse Stress Objective	**Function Integrated Risk**	**Modeling Disastrous Events**
Trading book	Methodological description, software-guided examples, application to Bank Alpha		
Banking book			
Liabilities			

1.4 SUMMARY

Opaque financial products, leveraged investments, and illiquidity acted as causes of instability during the 2007–09 crisis. A regulatory reform based on a new discipline of own funds, capital, and liquidity responded to the financial collapse. In this regard, stress testing was pinpointed by regulators across the world as a key instrument to assess bank resilience against adverse economic conditions. Thus an overview of the process was sketched in this initial chapter, through the introduction of Bank Alpha as a stress testing and risk integration illustrative example. From a stress testing perspective, we examined Bank Alpha's ratios by anticipating some of the issues to be investigated throughout the book. The need to enforce interdependent measures in risk management was highlighted by our outlining the key concepts of top-down and bottom-up risk integration approaches. Stress testing and risk integration outcomes showed Bank Alpha's capital weaknesses. In contrast, satisfactory regulatory liquidity indicators clashed with liquidity shortcomings pointed out by a fully integrated liquidity analysis. Finally, a book road map was depicted in an attempt to guide the reader during the stress testing and risk integration journey.

REFERENCES

Aas, K., Dimakos, X., Øksendal, A., 2007. Risk capital aggregation. Risk Manage. 9, 82–107.

Alessandri, P., Drehmann, M., 2010. An economic capital model integrating credit and interest rate risk in the banking book. J. Bank. Finance 34 (4), 730–742.

Atkinson, A.C., Riani, M., 2000. Robust Diagnostic Regression Analysis Springer, New York.

Atkinson, A.C., Riani, M., Cerioli, A., 2004. Exploring Multivariate Data with the Forward Search Springer, New York.

Bellini, T., 2013. Integrated bank risk modeling: a bottom-up statistical framework. Eur. J. Oper. Res. 230, 385–398.

BIS, 1988. International Convergence of Capital Measurement and Capital Standards. Bank for International Settlements, Basel.

BIS, 2006. Basel II International Convergence of Capital Measurement and Capital Standards: A Revised Framework. Bank for International Settlements, Basel.

BIS, 2011. Basel III: A global regulatory framework for more resilient banks and banking systems. Bank for International Settlements, Basel.

BOE, 2015. Stress testing the UK banking system: key elements of the 2015 stress test. Bank of England Publications, London.

Dimakos, X., Aas, K., 2004. Integrated risk modelling. Stat. Modell. 4 (4), 265–277.

Drehmann, M., Stringa, M., Sorensen, S., 2010. The integrated impact of credit and interest rate risk on banks: a dynamic framework and stress testing application. J. Bank. Finance 34, 713–729.

EBA, 2016. 2016 EU-wide stress test. Methodological note. European Banking Authority, London.

FRB, 2015. Comprehensive Capital Analysis and Review 2015: assessment framework and results Board of Governors of the Federal Reserve System, Washington, DC.

Grundke, P., 2009. Importance sampling for integrated market and credit portfolio models. Eur. J. Oper. Res. 194, 206–226.

Grundke, P., 2010. Top-down approaches for integrated risk management: how accurate are they?. Eur. J. Oper. Res. 203, 662–672.

Johansen, S., 1996. Likelihood-based Inference in Cointegrated Vector Autoregressive Models. Oxford University Press, Oxford.

Kretzschmar, G., McNeil, A., Kirchner, A., 2010. Integrated models of capital adequacy—why banks are undercapitalised. J. Bank. Finance 34 (12), 2838–2850.

Kuritzkes, A., Schuermann, T., Weiner, S., 2002. Risk measurement, risk management and capital adequacy in financial conglomerates. Working Paper 03-02. Wharton Financial Institutions Center, University of Pennsylvania.

Pesaran, M., Schuermann, T., Weiner, S., 2004. Modeling regional interdependencies using a global error-correcting macroeconometric model. J. Bus. Econ. Stat. 22, 129–162.

Chapter 2

Macroeconomic Scenario Analysis from a Bank Perspective

Chapter Outline

Time series analysis is vital for stress testing and risk integration. Two main perspectives can be followed to forecast the evolution of an economy. On the one hand, a structural approach aims to apply the economic theory to real data. However, it may suffer from some statistical deficiencies. On the other hand, a pure statistical method solves the latter issue but may lack a genuine economic interpretation. The practical difficulties in dealing with a structural method suggest we focus on a more marked econometric mindset.

Stress Testing and Risk Integration in Banks. http://dx.doi.org/10.1016/B978-0-12-803590-0.00002-3

As a first step, the study starts from the univariate autoregression (AR) and moving-average (MA) processes. Moreover, the Box-Jenkins analysis provides an accurate and parsimonious framework for the investigation of real time series. A practical implementation in MATLAB and R helps readers familiarize themselves with these modeling techniques.

Moving to the multivariate analysis, the traditional vector AR (VAR) and vector error-correction (VEC) models constitute the essential toolkit to deal with macroeconomic time series. Estimation, simulation, and forecast are components of a comprehensive process aimed at assessing the impact of external shocks on a bank business.

When interdependencies increase, models capturing international linkages become crucial. In this regard, the global VAR (GVAR) model facilitates a multicountry business stress test analysis as required in Bank Alpha's example. Additionally, a bank usually requires variables other than those included in a regulatory scenario to run its internal stress testing models. Therefore an enrichment process concludes this chapter by expediting the scrutiny of a wider set of coherent macroeconomic forecasts.

KEY ABBREVIATIONS AND SYMBOLS

X_t random variable at time t
x_t realized value of X_t
$\mathbf{x}_t$ vector of macroeconomic variables at time t
$\hat{\mathbf{x}}_t$ estimated vector of macroeconomic variables at time t
$\mathbf{x}_{t+h|t}$ h-step-ahead forecast, given the information at time t
$\mathbf{x}_{s,t}$ vector of macroeconomic variables for country s at time t
$\mathbf{x}_{s,t}^*$ vector of foreign countries macroeconomic variables for country s at time t
Δx_t $x_t - x_{t-1}$
$L^{(\cdot)}$ lag operator
ϵ_t white noise process (with normal error distribution)

2.1 INTRODUCTION

A macroeconomic scenario is at the very heart of a stress testing exercise. As a consequence, a bank needs to investigate how macroeconomic movements affect its own business. Time series analysis is a key tool within this risk management process. A debate characterizes the use of structural versus econometric models to explain macroeconomic fluctuations. One of the key issues related to a structural model (i.e., dynamic stochastic general equilibrium), as introduced by Kydland and Prescott (1982), is parameter integrity. In contrast, the flexibility and reliability of statistical models (e.g., VAR and VEC) suggests their use for stress testing and risk integration. Indeed, this latter perspective is followed throughout this book to build a bridge between macroeconomic scenarios and internal risk systems.

Three main goals inspire this chapter. Firstly, one needs to become accustomed to the commonest econometric tools to fit macroeconomic time series and make projections. The second objective is to grasp scenario generation techniques widely used for risk integration. A bottom-up framework conventionally relies on a series of scenarios from which a loss distribution can be derived. Therefore model parameters do not feed only a specific path, but a wide range of alternative outlines feeds a more comprehensive assessment system. Undoubtedly, one of the key issues a risk manager faces when dealing with a stress test is to include variables in a scenario other than those provided by the regulator. The investigation of an enrichment method in line with Doan et al. (1984) is the third target pursued in this chapter.

As a corollary of the above, the reader needs to be aware that all these models are ingredients of a more extended system where a direct or indirect feed is required. Indeed, macroeconomic variables directly feed a stress test framework by affecting the term structure of interest rates, market risk parameters (e.g., volatility), credit risk factors (e.g., probability of default), operational losses, and balance sheet projections. These components indirectly nourish the projection of profit and loss, risk-weighted assets, and capital and liquidity ratios.

The chapter is organized as follows. The brief introduction to univariate AR and MA time series in Section 2.2 promptly leaves the floor to multivariate studies (Lutkepohl, 2005). The bulk of the analysis proposed in Section 2.3 dwells on VAR and VEC. Section 2.4 focuses on GVAR models (Pesaran et al., 2004) while Section 2.5 investigates how to design and enrich macroeconomic scenarios.

From a toolkit perspective, practical examples in MATLAB and R help the reader to grasp how to estimate, diagnose, and project univariate and multivariate time series and perform practical time series analysis. The appendix shows how robust techniques enhance traditional estimation methods (Atkinson et al., 2004).

2.2 AUTOREGRESSION AND MOVING-AVERAGE MODELING

A time series is a set of observations x_t, each one being recorded at a specific time t. In a discrete time series, observations are captured at fixed time intervals. In contrast, in continuous time series, observations are recorded continuously over a time interval (e.g., the interval is [0, 1]). It is natural to suppose that each observation x_t is a realized value of a certain random variable X_t. In this section, univariate time series are scrutinized by our focusing on AR, MA, and AR and MA (ARMA) models. A brief description of the AR integrated MA (ARIMA) model paves the way for the concept of stationarity explored through the multivariate cointegration analysis (see Section 2.3). A series of exercises at the end of the chapter will help readers familiarize themselves with the univariate time series. These studies are preparatory to the more comprehensive multivariate analysis.

2.2.1 AR(p) Analysis

A useful starting point is the AR process. This simple process introduces the key idea underlying all subsequent and more complex ones. In what follows, the typical first-order (i.e., one time lag) AR(1) formulation is shown:

$$x_t = c + \phi x_{t-1} + \epsilon_t, \tag{2.1}$$

where x_t represents the current period value of the process, x_{t-1} is the previous period value, c is a constant, and ϵ_t is a white noise component. Errors are assumed to be normally distributed, with $\mathbb{E}(\epsilon_t) = 0$, $\mathbb{E}(\epsilon_t^2) = \sigma^2$, and $\mathbb{E}(\epsilon_t \epsilon_\tau) = 0$ for $t \neq \tau$.

The path of this process critically depends on the value of ϕ. If $|\phi| \geq 1$, the process is nonstationary. In particular, $|\phi| > 1$ implies that the process grows without bound. $|\phi| = 1$ drives a unit root process.

From a formal point of view, it is useful to rewrite Eq. (2.1) in terms of the lag operator (L) as follows:

$$(1 - \phi L) x_t = c + \epsilon_t. \tag{2.2}$$

Then the following infinite sum of past error (with decaying weights) is obtained by substituting Eq. (2.1) into Eq. (2.2):

$$x_t = (c + \epsilon_t) + \phi (c + \epsilon_{t-1}) + \phi^2 (c + \epsilon_{t-2}) + \cdots \tag{2.3}$$

$$= \frac{c}{1 - \phi} + \epsilon_t + \phi \epsilon_{t-1} + \phi^2 \epsilon_{t-2} + \cdots .$$

The AR(p) generalization of the AR(1) process of Eq. (2.1) is detailed as follows:

$$x_t = c + \phi_1 x_{t-1} + \phi_2 x_{t-2} + \cdots + \phi_p x_{t-p} + \epsilon_t. \tag{2.4}$$

According to the lag operator representation presented above, Eq. (2.2) can be rewritten for the AR(p) process as:

$$\left(1 - \phi_1 L - \phi_2 L^2 - \cdots - \phi_p L^p \right) x_t = c + \epsilon_t. \tag{2.5}$$

In line with the assumption of normally distributed errors, Eq. (2.4) can be consistently estimated by use of the ordinary least squares method. In this regard, Exercise 2.1 describes how to simulate and estimate an AR(2) process with the use of MATLAB. Exercise 2.2 focuses on the ordinary least squares estimation of the same AR(2) process introduced in Exercise 2.1.

The next section introduces MA processes.

2.2.2 MA(q) Analysis

Eq. (2.3) showed that an AR process can be represented as an MA of contemporaneous and past shocks. The simplest MA process relies on one lag as detailed below:

$$x_t = \eta + \epsilon_t + \theta \epsilon_{t-1}, \tag{2.6}$$

where ϵ_t is a white noise process and η and θ are constants.

Starting from the MA(1) process, an MA(q) process is easy to describe as shown in the next equation:

$$x_t = \eta + \epsilon_t + \theta_1 \epsilon_{t-1} + \cdots + \theta_q \epsilon_{t-q}. \tag{2.7}$$

In terms of the lag operator, Eq. (2.7) can be written in the following way:

$$x_t - \eta = \epsilon_t + \theta_1 \epsilon_{t-1} + \cdots + \theta_q \epsilon_{t-q} \tag{2.8}$$
$$= (1 + \theta_1 L + \cdots + \theta_q L^q)\epsilon_t.$$

When one is analyzing real time series, a combination of AR and MA components is used. Hence, ARMA processes are a useful paradigm, as outlined in the next section.

2.2.3 ARMA(p, q) Analysis

When pure AR and MA processes are not capable of capturing real time series paths, an ARMA process may represent a solution. An ARMA(p, q) process is defined as follows:

$$x_t = c + \phi_1 x_{t-1} + \cdots + \phi_p x_{t-p} + \epsilon_t + \theta_1 \epsilon_{t-1} + \cdots + \theta_q \epsilon_{t-q}. \tag{2.9}$$

Considering the lag operator, we can write Eq. (2.9) as follows:

$$x_t = \frac{c}{1 - \phi_1 L - \cdots - \phi_p L^p} + \frac{1 + \theta_1 L + \cdots + \theta_q L^q}{1 - \phi_1 L - \cdots - \phi_p L^p}\epsilon_t = \mu + \psi(L)\epsilon_t. \tag{2.10}$$

Exercise 2.3 shows how to generate an ARMA(1,1) process and fit it with the software R.

It is worth mentioning that the above ARMA(p, q) representation may be used when the underlying time series is stationary. What does stationarity mean in practice? Let us assume that a time series x_t is a realization of a deterministic trend and a stochastic component:

$$x_t = TD_t + v_t, \tag{2.11}$$

where TD_t is a deterministic trend ($TD_t = \beta_1 + \beta_2 t$) and v_t represents the stochastic component $\phi(L)v_t = \theta(L)\epsilon_t$ with ϵ_t independent and identically distributed. Thus the two following cases may be represented:

- **All roots outside the unit circle.** When all roots of the autoregressive polynomial lie outside the unit circle, x_t is stationary around a deterministic trend. This model is termed an *integrated model of order 0, I(0)*.
- **All roots outside, but one on the unit circle.** In this case, $\Delta v_t = (1 - L)v_t$ is stationary around a constant mean. The difference stationary model is referred to as an *integrated model of order 1, I(1)*.

An ARMA(p, q) model can be fitted to the differenced time series. This model is referred as *ARIMA(p, d, q)*, where d stands for the order of integration (i.e., the number of times the original series must be differenced until a stationary one is obtained). The stochastic component can be decomposed into cycle and trend components to better understand the process under analysis. The cyclical component is assumed to be a mean-stationary process, whereas all random shocks are captured by the stochastic component. Thus the generating process may be split into deterministic trend, stochastic trend, and cyclical components as detailed below:

- **Trend stationary model.** In this case the stochastic trend is zero and the cyclical component is equal to the ARMA(p, q) model.
- **Difference-stationary model.** The autoregressive polynomial contains a unit root that can be factored out as $\phi(L) = (1 - L)\phi^*(L)$, where the roots of the polynomial $\phi^*(L)$ are outside the unit circle. Then it is possible to represent Δv_t as an MA process.

In statistics, the random walk process has been studied as a prototype of a unit root process:

$$x_t = x_{t-1} + \epsilon_t = x_0 + \sum_{r=1}^{t} \epsilon_r, \tag{2.12}$$

where ϵ_t is white noise. It is evident that the random walk process is nonstationary because its variance increases with time. The random walk with drift is as follows:

$$x_t = \mu + x_{t-1} + \epsilon_t = x_0 + \mu t + \sum_{r=1}^{t} \epsilon_r, \tag{2.13}$$

where μ is a constant term. The sign of the drift, μ, causes the series to go upward or downward. The absolute value of this constant term affects the steepness of the series.

Time series stationarity is usually verified through the so-called Dickey-Fuller test (Dickey and Fuller, 1981). Many variations and hypotheses are conceivable within this framework. Hereafter, the focus is on the simplest version of an AR(1) process and the question underlying the test is as follows: Is the underlying process a random walk (i.e., $\phi = 1$) or is it stationary (i.e., $\phi < 1$)? Formally, the test works as shown below:

$$x_t = \phi x_{t-1} + \epsilon_t, \tag{2.14}$$

$$H_0 : \phi = 1, \tag{2.15}$$

$$H_1 : \phi < 1, \tag{2.16}$$

where ϵ_t is normally distributed.

The model can also be written in (first) differences. Thus if we subtract x_{t-1} from both sides of the equation, write $\Delta x_t = x_t - x_{t-1}$, and introduce $\alpha = (\phi - 1)$, the following holds:

$$\Delta x_t = \alpha x_{t-1} + \epsilon_t, \tag{2.17}$$

$$H_0 : \alpha = 0, \tag{2.18}$$

$$H_1 : \alpha < 0, \tag{2.19}$$

where now the hypothesis is in terms of α. The natural way of testing whether $\alpha = 0$ is to regress Δx_t on x_{t-1} and run a T test having the following ratio form $\frac{\hat{\alpha}}{\text{est.(s.e. of } \hat{\alpha})}$. In this setting, $\hat{\alpha}$ is the least squares estimate of α and the null is rejected for large negative values. If the null were $\alpha = -0.1$ (i.e., ϕ=0.9) and the process were stationary, then the Student T distribution associated with the null hypothesis would be $N(0, 1)$ in large samples. This fits with the ordinary regression case where the Student T distribution tends to be the standard normal when the degrees of freedom tend to infinity. However, when $\phi \geq 1$, the T statistic is not $N(0, 1)$ in large samples. Dickey and Fuller (1981) worked out the distribution for $\phi = 1$ and tabulated it.

The basic Dickey-Fuller test exists in variants, for example:

$$\Delta x_t = \mu + \alpha x_{t-1} + \epsilon_t, \tag{2.20}$$

$$\Delta x_t = \mu + \alpha x_{t-1} + \beta t + \epsilon_t. \tag{2.21}$$

In the *augmented* Dickey-Fuller (ADF) test, additional terms are included in the regression to cover the possibility of higher order AR:

$$\Delta x_t = \mu + \alpha x_{t-1} + \beta_1 \Delta x_{t-1} + \cdots + \beta_p \Delta x_{t-p} + \epsilon_t, \tag{2.22}$$

where the test is about $\alpha = 0$ as above.

When written in terms of $\Delta x_t = x_t - x_{t-1}$, the random walk process described in Eq. (2.12) becomes

$$\Delta x_t = \epsilon_t, \tag{2.23}$$

where, as in Eq. (2.16), errors are normally distributed. This first difference process is a white noise. Such a process is said to be integrated of order 1 and is denoted as $I(1)$. When it is necessary to difference twice to achieve stationarity, then the series is called $I(2)$, while a stationary series is $I(0)$. This concept is used extensively in the following sections, starting with the Box-Jenkins analysis.

2.2.4 Box-Jenkins Time Series Analysis

Some important questions arise when one is fitting a model to real data. What are the appropriate values for (p, q) when one is using an ARMA model? How should estimates correspond to unknown parameter values? Box and Jenkins (1976) proposed the following three-step framework (i.e., specification, estimation and selection, diagnostic) through which to answer these questions:

- **Specification**[1]. This step involves determining the order of the model to capture data dynamics. Graphical procedures such as plotting the time

---

1. The term identification is commonly used in econometrics, instead of specification. Notwithstanding, specification is used to avoid misunderstanding related to the identification process to constrain VAR and VEC models.

series, autocorrelation function, and partial autocorrelation function help the analysis. Nonetheless, estimation and diagnostic steps need to be performed to reach the final conclusion on model specification.

- **Estimation and selection.** This step relies on the estimation of different models' parameters. As a consequence, models are compared according to information criteria such as the Akaike information criterion (AIC), the Bayesian information criterion (BIC), and the Hannan-Quinn information criterion (Lutkepohl, 2005).
- **Diagnostic.** The last step of the process involves determining whether the specified and estimated model is adequate. Notably, residual diagnostics are performed.

It is worth highlighting that these three steps are strictly related to each other. Thus the final specification depends on estimation and selection as well as diagnostic outcomes.

Example 2.1 shows how the Box-Jenkins framework works in practice. The MATLAB Econometric Toolbox is used to perform the analysis of UK inflation, computed as consumer price index (CPI) log difference (i.e., $\ln(CPI_t) - \ln(CPI_{t-1})$) from 2000 to 2013. This time series is part of the GVAR toolbox available from https://sites.google.com/site/gvarmodelling/gvar-toolbox.

Example 2.1 Box-Jenkins Analysis

Let us consider the UK inflation time series from 2000 to 2013. Specification, estimation and selection, and diagnostic are performed interactively. The goal is to figure out the model that most parsimoniously fits the data (i.e., avoids overparametrization) and does not violate the theoretical assumptions on which the fitting process relies.

The first step of the process is to upload the time series and check for stationarity via the ADF test as detailed below:

```
macv=xlsread('Chap2UKmacvar.xlsx');
infl=macv(2:end,14);
% Stationarity ADF test
[h,pvalue]=adftest(infl);
```

The ADF test suggests the rejection of the unit root null in favor of the alternative, meaning that the process is stationary ($p = 0.0437$).

The analysis continues by graphically inspecting the time series and estimating alternative candidate models. The following ARIMA MATLAB function is used to specify the model to be estimated. This function has three inputs: the order of the AR, integration and MA. Given the ADF test result, the integration parameter is set equal to 0. The following models are considered: ARIMA(1,0,0), ARIMA(0,0,1), and ARIMA(1,0,1).

The ARIMA(1,0,0) MATLAB code to fit the time series is as follows:

```
% 1. Specification
Mdl = arima(1,0,0);
```

Example 2.1 Box-Jenkins Analysis—cont'd

```
% 2. Selection
% 2.1 Estimation
[inflEstMdl inflEstMdlParamCov infllogL]=estimate(Mdl,infl);
```

The choice of the lag order needs to be supported by AIC and BIC analysis as detailed below:

```
% 2.2 Information criteria
[aic,bic] = aicbic(infllogL,2,size(infl,1));
```

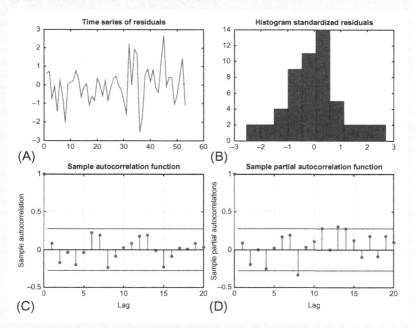

(A) **(B)** **(C)** **(D)**

FIG. 2.1 Box-Jenkins residual diagnostics for the ARIMA(1,0,0) model.

As part of the Box-Jenkins procedure, residuals are tested (see Fig. 2.1). Firstly, normality is checked (normality test), then autocorrelation is verified (Ljung-Box autocorrelation test). The function *lillietest* returns a test decision for the null hypothesis that the data in the vector come from a distribution in the normal family. The alternative is that they do not come from such a distribution. The result is 1 if the test rejects the null hypothesis, and 0 otherwise (i.e., when residuals are normally distributed, hNorm0 = 0). The Ljung-Box test for autocorrelation is conducted through the function *lbqtest*. It returns a logical value, with the rejection decision indicating the rejection of the no residual autocorrelation null hypothesis in favor of the alternative. In contrast, hLBQ0 = 0 indicates failure to reject the no residual autocorrelation null hypothesis (i.e., residuals are not autocorrelated when hLBQ0 = 0).

(Continued)

Example 2.1 Box-Jenkins Analysis—cont'd

```
% 3. Model checkDiagnostics
% Residuals
[resinflFit] = infer(inflEstMdl,infl);
% 3.1 Normality
[hNorm0,pNorm0] = lillietest(resinflFit, 'Alpha',0.01);
% 3.2 Ljung-Box Q-test
[hLBQ0,pValueLBQ0] = lbqtest(resinflFit,'Lags',[5,10],...
    'Alpha',0.01);
```

Table 2.1 summarizes the key statistics for all the candidate models. The comparison of the AIC and BIC columns highlights that the ARIMA(0,0,1) model should be preferred to the others. However, diagnostic checking shows that the ARIMA(1,0,0) model performs quite well in terms of both normality and nonautocorrelation of residuals. Potential overparametrization issues would characterize the use of the ARIMA(1,0,1) model.

TABLE 2.1 Box-Jenkins Diagnostics: UK Inflation Time Series

	AIC	BIC	pNorm	pValueLBQ	
				Lag 5	Lag 10
ARIMA(1,0,0)	−456.7028	−452.7622	0.3340	0.4637	0.1568
ARIMA(0,0,1)	−460.2066	−456.2660	0.3049	0.8660	0.5856
ARIMA(1,0,1)	−459.0596	−453.1488	0.2274	0.8515	0.6890

AIC, Akaike information criterion; BIC, Bayesian information criterion; pNorm, pvalue normality test; pValueLBQ, pvalue Ljung-Box test.

This brief introduction to univariate time series paves the way to a more comprehensive multivariate analysis on which stress testing and risk integration rely. The next section focuses on VAR and VEC models.

2.3 VECTOR AUTOREGRESSION AND VECTOR ERROR-CORRECTION MODELING

A series of approaches have been proposed in the literature to describe the behavior of economic and financial time series (Lutkepohl, 2005). The VAR model has proven to be a flexible tool in capturing statistical patterns and ductile in representing theory-based relationships. When we are dealing with integrated time series, the VEC model serves as a model archetype, as detailed in the following sections.

2.3.1 Vector Autoregression and Vector Error-Correction Analysis

The VAR model is a natural extension of the univariate AR model to dynamic multivariate time series. Let $\mathbf{x}_t = (x_{1,t}, \ldots, x_{p,t})'$ denote a $p \times 1$ vector of time series variables. The basic k-lag VAR(k) model has the following form:

$$\mathbf{x}_t = \mathbf{c} + \mathbf{\Phi}_1 \mathbf{x}_{t-1} + \cdots + \mathbf{\Phi}_k \mathbf{x}_{t-k} + \boldsymbol{\epsilon}_t, \qquad (2.24)$$

$$\underbrace{\begin{bmatrix} x_{1,t} \\ \vdots \\ x_{p,t} \end{bmatrix}}_{x_t} = \underbrace{\begin{bmatrix} c_1 \\ \vdots \\ c_p \end{bmatrix}}_{c} + \underbrace{\begin{bmatrix} \phi_{1_{1,1}} & \cdots & \phi_{1_{1,p}} \\ \vdots & \ddots & \vdots \\ \phi_{1_{p,1}} & \cdots & \phi_{1_{p,p}} \end{bmatrix}}_{\Phi_1} \underbrace{\begin{bmatrix} x_{1,t-1} \\ \vdots \\ x_{p,t-1} \end{bmatrix}}_{x_{t-1}} + \cdots$$

$$+ \underbrace{\begin{bmatrix} \phi_{k_{1,1}} & \cdots & \phi_{k_{1,p}} \\ \vdots & \ddots & \vdots \\ \phi_{k_{p,1}} & \cdots & \phi_{k_{p,p}} \end{bmatrix}}_{\Phi_k} \underbrace{\begin{bmatrix} x_{1,t-k} \\ \vdots \\ x_{p,t-k} \end{bmatrix}}_{x_{t-k}} + \underbrace{\begin{bmatrix} \epsilon_{1,t} \\ \vdots \\ \epsilon_{p,t} \end{bmatrix}}_{\epsilon_t},$$

where $\Phi_{1,\ldots,k}$ are $p \times p$ matrices of coefficients and ϵ_t is a $p \times 1$ zero mean white noise vector process (serially uncorrelated or independent) with time-invariant covariance matrix Σ. The VAR(k) model has the following form in lag operator notation:

$$\Phi(L)x_t = c + \epsilon_t, \tag{2.25}$$

where $\Phi(L) = I_p - \Phi_1 L - \cdots - \Phi_k L^k$. The VAR($k$) model is stable if the eigenvalues of the companion matrix

$$C = \begin{bmatrix} \Phi_1 & \Phi_2 & \cdots & \Phi_p \\ I_p & 0 & \cdots & 0 \\ 0 & \ddots & 0 & \vdots \\ 0 & 0 & I_p & 0 \end{bmatrix} \tag{2.26}$$

have modulus less than 1.

Example 2.2 is an easy introduction to multivariate time series modeling. It highlights how to generate a VAR(1) process and estimate its parameters through the MATLAB Econometric Toolbox. It also serves the purpose of illustrating the key features of the conditional forecasting outlined in Section 2.5.

Example 2.2 Simulation and Estimation of a VAR(1) Model

Let us consider the VAR(1) model

$$\begin{bmatrix} x_{1,t} \\ x_{2,t} \end{bmatrix} = \begin{bmatrix} 0.2 \\ 0.2 \end{bmatrix} + \begin{bmatrix} 0.3 & 0.2 \\ 0.2 & 0.3 \end{bmatrix} \begin{bmatrix} x_{1,t-1} \\ x_{2,t-1} \end{bmatrix} + \begin{bmatrix} \epsilon_{1,t} \\ \epsilon_{2,t} \end{bmatrix},$$

where the covariance matrix Σ is as follows:

$$\begin{bmatrix} 1.0 & 0.8 \\ 0.8 & 1.0 \end{bmatrix}.$$

A MATLAB code is implemented by our pursuing the following goals:
- simulate 50 realizations of the VAR(1) model detailed above. It is worth noting that the number of observations required in time series analysis is

(Continued)

Example 2.2 Simulation and Estimation of a VAR(1) Model—cont'd

usually greater. However, in risk management practice, analysts often face data availability issues. For this reason a small number of observation is considered in this example;

● check time series stationarity;
● estimate the simulated time series model parameters;
● perform diagnostic checks.

The first step of the simulation process is to set the model parameters by use of the function *vgxset*. Then the *vgxsim* function is used to generate 50 realizations of the VAR(1) time series as follows:

```
% 1. Simulation
Spec = vgxset(...
'a', [0.2; 0.2], ...
'AR', {[0.3, 0.2; 0.2, 0.3]}, ...
'Q', [1, 0.8; 0.8, 1]);
numObs=50;
seed=12;rng(seed); % set seed
VAR1model = vgxsim(Spec,numObs);
```

The eigenvalues of the companion matrix as well as the ADF test show that the time series are stationary as shown below.

```
% 2. Stationarity check
% Eigenvalues
eigenv=eig([0.3, 0.2; 0.2, 0.3]);
% Output
% 0.1000 0.5000
% ADF test
[hV1,pvalueV1]=adftest(VAR1model(:,1));
[hV2,pvalueV2]=adftest(VAR1model(:,2));
% Output
% hV1 = 1  pvalueV1 =    1.0000e-03
% hV2 = 1  pvalueV2 =    1.0000e-03
```

The model is fitted by use of the function *vgxvarx*, and parameter estimates are shown in the output table with their corresponding *T* statistics. Fig. 2.2 compares simulated and fitted time series. Diagnostic checks indicate a failure to reject the no residual autocorrelation null hypothesis at the 1% confidence level. Similarly, the normality distribution null cannot be rejected.

```
% 3. Estimation of the VAR(1) process
SpecVAR=vgxset('n',2,'nAR',1, 'Constant',true);
[EstSpec,EstStdErrors,LLF,W]= vgxvarx(SpecVAR,VAR1model);
vgxdisp(EstSpec, EstStdErrors);
% Output
%  Model  : 2-D VAR(1) with Additive Constant
%     Conditional mean is AR-stable and is MA-invertible
%     Standard errors without DoF adjustment (max.likelih.)
```

Example 2.2 Simulation and Estimation of a VAR(1) Model—cont'd

```
%      Parameter            Value            Std. Error       t-Statistic
%
% _ _ _ _ _ _           _ _ _ _ _           _ _ _ _ _          _ _ _ _ _
%           a(1)           0.25496            0.157005            1.6239
%           a(2)           0.29826            0.160368            1.85985
%   AR(1)(1,1)           0.224719            0.246773            0.910628
%        (1,2)           0.252047            0.240863            1.04643
%        (2,1)           0.294353            0.252059            1.16779
%        (2,2)           0.180505            0.246023            0.733691
%        Q(1,1)           0.959693
%        Q(2,1)           0.805833
%        Q(2,2)           1.00125
% 4. Diagnostic
% Ljung-Box Q-test
res=VAR1model-W;
[hV1,pV1] = lbqtest(res(:,1),'Lags',[5,10], 'Alpha', 0.01);
[hV2,pV2] = lbqtest(res(:,2),'Lags',[5,10], 'Alpha', 0.01);
% hV1 = 0  0;  pV1 = 0.0106      0.0112;
% hV2 = 0  0;  pV2 = 0.0109      0.0142;
% Normality
[hnV1,pnV1] = lillietest(res(:,1), 'Alpha', 0.01);
[hnV2,pnV2] = lillietest(res(:,2), 'Alpha', 0.01);
% hnV1 =  0; pnV1 =  0.3826;
% hnV2 =  0; pnV2 =  0.1480;
```

Fig. 2.2 outlines a comparison between the simulated $x_{1,t}$ and $x_{2,t}$ and the fitted $\hat{x}_{1,t}$ and $\hat{x}_{2,t}$.

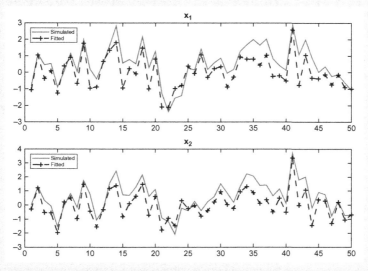

FIG. 2.2 Simulated VAR(1) $x_{1,t}$ and $x_{2,t}$ (*plain line*) and the fitted model (*dotted line*).

When we are dealing with real time series, the stationarity assumption does not necessary hold. In such cases, cointegration is used.

Johansen (1988) highlighted that a $p \times 1$ vector of variables $\mathbf{x}_t = (x_{1,t}, \ldots x_{p,t})'$, $t = 1, \ldots, T$, is said to be cointegrated if at least one nonzero $p \times 1$ vector $\boldsymbol{\beta}_l$ exists such that $\boldsymbol{\beta}_l' \mathbf{x}_t$ is trend stationary. In this case, $\boldsymbol{\beta}_l$ is called a *cointegrating vector*. If r such linearly independent vectors exist, $\mathbf{x}_t$ is cointegrated with cointegrating rank r. Starting from the $p \times r$ matrix of cointegrating vectors $\boldsymbol{\beta} = (\boldsymbol{\beta}_1, \ldots, \boldsymbol{\beta}_r)$, the r elements of the vector $\boldsymbol{\beta}' \mathbf{x}_t$ are trend stationary and $\boldsymbol{\beta}$ is called the *cointegrating matrix*.

According to Johansen (1996), the cointegrated VAR can be represented in the form of VEC as follows:

$$\Delta \mathbf{x}_t = \boldsymbol{\Pi} \mathbf{x}_{t-1} + \boldsymbol{\Gamma}_1 \Delta \mathbf{x}_{t-1} + \cdots + \boldsymbol{\Gamma}_{k-1} \Delta \mathbf{x}_{t-k+1} + \boldsymbol{\epsilon}_t, \qquad (2.27)$$

$$\underbrace{\begin{bmatrix} \Delta x_{1,t} \\ \vdots \\ \Delta x_{p,t} \end{bmatrix}}_{\Delta \mathbf{x}_t} = \underbrace{\begin{bmatrix} \pi_{1,1} & \cdots & \pi_{1,p} \\ \vdots & \ddots & \vdots \\ \pi_{p,1} & \cdots & \pi_{p,p} \end{bmatrix}}_{\boldsymbol{\Pi}} \underbrace{\begin{bmatrix} x_{1,t-1} \\ \vdots \\ x_{p,t-1} \end{bmatrix}}_{\mathbf{x}_{t-1}} + \underbrace{\begin{bmatrix} \gamma 1_{1,1} & \cdots & \gamma 1_{1,p} \\ \vdots & \ddots & \vdots \\ \gamma 1_{p,1} & \cdots & \gamma 1_{p,p} \end{bmatrix}}_{\boldsymbol{\Gamma}_1} \underbrace{\begin{bmatrix} \Delta x_{1,t-1} \\ \vdots \\ \Delta x_{p,t-1} \end{bmatrix}}_{\Delta \mathbf{x}_{t-1}}$$

$$+ \cdots + \underbrace{\begin{bmatrix} \gamma_{k-1_{1,1}} & \cdots & \gamma_{k-1_{1,p}} \\ \vdots & \ddots & \vdots \\ \gamma_{k-1_{p,1}} & \cdots & \gamma_{k-1_{p,p}} \end{bmatrix}}_{\boldsymbol{\Gamma}_{k-1}} \underbrace{\begin{bmatrix} \Delta x_{1,t-k+1} \\ \vdots \\ \Delta x_{p,t-k+1} \end{bmatrix}}_{\Delta \mathbf{x}_{t-k+1}} + \underbrace{\begin{bmatrix} \epsilon_{1,t} \\ \vdots \\ \epsilon_{p,t} \end{bmatrix}}_{\boldsymbol{\epsilon}_t},$$

where $\mathbf{x}_t = (x_{1,t} \ldots x_{p,t})'$ is the $p \times 1$ vector of variables and Δ is the first difference operator, (i.e., $\Delta \mathbf{x}_t = \mathbf{x}_t - \mathbf{x}_{t-1}$). In the case where $\boldsymbol{\Pi}$ has a reduced rank, the following equation holds: $\boldsymbol{\Pi} = \boldsymbol{\alpha} \boldsymbol{\beta}'$, where $\boldsymbol{\alpha}$ and $\boldsymbol{\beta}$ are $p \times r$ matrices, $r \leq p$. $\boldsymbol{\beta}$ is the cointegrating matrix described above. $\boldsymbol{\Gamma}_1, \ldots, \boldsymbol{\Gamma}_{k-1}$ are $p \times p$ parameter matrices. Errors are assumed to be normally distributed $\boldsymbol{\epsilon}_t \sim N(\mathbf{0}, \boldsymbol{\Sigma})$, where $\boldsymbol{\Sigma}$ represents the $p \times p$ covariance matrix. In addition, one could consider a vector $\mathbf{D}_t = (D_{1,t}, \ldots, D_{g,t}, \mu_0)'$ containing g (binary) dummies and a constant. The following three cases need to be considered when one is assessing the rank of $\boldsymbol{\Pi}$:

$$\text{rank}(\boldsymbol{\Pi}) = p,$$

$$\text{rank}(\boldsymbol{\Pi}) = 0,$$

$$0 < \text{rank}(\boldsymbol{\Pi}) \leq p.$$

In the first case, all p linearly independent combinations must be stationary. In the second case, no linear combination exists to make $\boldsymbol{\Pi} \mathbf{x}_t$ stationary, except for the trivial solution. The most important case is the third one, where the rank of $\boldsymbol{\Pi}$ is greater than 0, but lower than p. As stated earlier, when the matrix has no full rank, it can be decomposed into the product of two $p \times r$ matrices $\boldsymbol{\alpha}$ and $\boldsymbol{\beta}$ such that $\boldsymbol{\alpha} \boldsymbol{\beta}'$. Hence $\boldsymbol{\alpha} \boldsymbol{\beta}' \mathbf{x}_t$ is stationary. The r linear independent columns of $\boldsymbol{\beta}$

are the cointegrating vectors. Each vector represents one long-term relationship between the individual series of $\mathbf{x}_t$.

The parameters of matrices $\boldsymbol{\alpha}$ and $\boldsymbol{\beta}$ are undefined because any nonsingular matrix $\boldsymbol{\Theta}$ would yield $\boldsymbol{\alpha}\boldsymbol{\Theta}\left(\boldsymbol{\beta}\boldsymbol{\Theta}^{-1}\right)' = \boldsymbol{\Pi}$. This implies that only the cointegration space spanned by $\boldsymbol{\beta}$ can be determined. The obvious solution is to normalize one element of $\boldsymbol{\beta}$ to 1. The reader may refer to Juselius (2006) for an interesting discussion on identification. The elements of $\boldsymbol{\alpha}$ determine the speed of adjustment to the long-term equilibrium.

Example 2.3 outlines how to deal with cointegration modeling. As per Example 2.1, the UK time series are considered as uploaded from the GVAR toolbox database covering the period from 2000 to 2013.

Example 2.3 VEC Analysis

Let us consider the following UK time series: nominal gross domestic product (GDP), consumer price index (CPI), nominal equity price index (EQ), exchange rate in US dollars (ER), nominal short-term interest rate per annum in percent (R^{ST}), and nominal long-term interest rate per annum in percent (R^{LT}). The analysis covers the period from 2000 to 2013. Suitable variable transformations are summarized in Table 2.2.

TABLE 2.2 UK Variable Transformation Used for Vector Error-Correction Analysis

Descriptions	Symbols and Analytical Formulas
Real output	$y_t = \ln(GDP_t/CPI_t)$
ln CPI	$\pi_t = \ln(CPI_t)$
Real equity price	$eq_t = \ln(EQ_t/CPI_t)$
Real exchange rate	$er_t = \ln(ER_t/CPI_t)$
Short-term interest rate	$r_t^{ST} = 0.25\ln(1 + R_t^{ST}/100)$
Long-term interest rate	$r_t^{LT} = 0.25\ln(1 + R_t^{LT}/100)$

The analysis is performed through the following steps by use of the MATLAB Econometric Toolbox:
- check for stationarity through the ADF test;
- specification;
- estimation and selection;
- diagnostic.

```
% File upload
mac=xlsread('Chap2UKmacvar.xlsx');
macv=mac(2:end,7:12);
```

(Continued)

Example 2.3 VEC Analysis—cont'd

Fig. 2.3 shows the dynamics of all the UK variables under analysis during the period from 2000 to 2013.

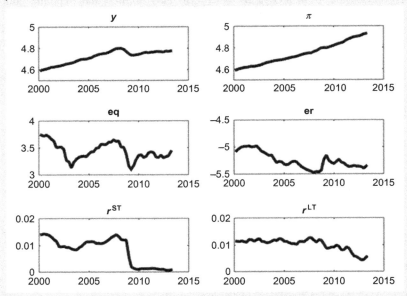

FIG. 2.3 UK macroeconomic time series analysis from 2000 to 2013.

The ADF test verifies time series stationarity. According to this test, Table 2.3 shows that the unit root null hypothesis cannot be rejected for the original time series. In contrast, the unit root hypothesis is rejected for all first differenced time series.

```
% 1. Check for stationarity (ADF test) original time series
[hGDP,pvalueGDP]=adftest(macv(:,1));
[hDP,pvalueCPI]=adftest(macv(:,2));
[hEQ,pvalueEQ]=adftest(macv(:,3));
[hER,pvalueER]=adftest(macv(:,4));
[hRS,pvalueRS]=adftest(macv(:,5));
[hRL,pvalueRL]=adftest(macv(:,6));
% 1.1 ADF test on first difference transformation
GDPd1= diff(macv(1:end,1)); [hGDPd1,pvalueGDPd1]=adftest
                                                  (GDPd1);
CPId1= diff(macv(1:end,2)); [hCPId1,pvalueCPId1]=adftest
                                                  (CPId1);
EQd1= diff(macv(1:end,3)); [hEQd1,pvalueEQd1]=adftest(EQd1);
ERd1= diff(macv(1:end,4)); [hERd1,pvalueERd1]=adftest(ERd1);
RSd1= diff(macv(1:end,5)); [hRSd1,pvalueRSd1]=adftest(RSd1);
RLd1= diff(macv(1:end,6)); [hRLd1,pvalueRLd1]=adftest(RLd1);
```

Example 2.3 VEC Analysis—cont'd

TABLE 2.3 Augmented Dickey-Fuller Test for Original Variables and Their First Differences

	Original Variable pvalue	Δ Variable pvalue
y_t	0.9990	0.0067
π_t	0.9990	0.0381
eq	0.3824	1.0000-03
er	0.8923	1.0000e-03
r^{ST}	0.0791	1.0000e-03
r^{ST}	0.2024	1.0000e-03

In line with the univariate time series, Box-Jenkins analysis applies. Model specification, estimation, and selection occur in conjunction with diagnostic checking. Firstly, the VEC model with one lag (and constant) is analyzed. The Johansen trace test shows a rank $r = 1$ at the 5% confidence level and $r = 2$ at the 10% confidence level. When the number of lags is increased to estimate a VEC(2), the Johansen trace test highlights $r = 0$. The one-lag model is then investigated as detailed below.

```
% 2. Specification
mod='H1*';
lagCVAR=1;
% 3. Estimation-selection
[h,pValue,stat,cValue,mles] = ...
jcitest(macv,'model',mod,'lags',lagCVAR, 'test', 'trace');
% Output
% Results Summary (Test 1)
% Data: macv
% Effective sample size: 51
% Model: H1*
% Lags: 1
% Statistic: trace
% Significance level: 0.05
% _ _ _ _ _ _ _ _ _ _ _ _ _ _ _ _ _ _ _ _ _
% r  h  stat       cValue     pValue    eigVal
% _ _ _ _ _ _ _ _ _ _ _ _ _ _ _ _ _ _ _ _ _
% 0  1  125.7773   103.8476   0.0010    0.5950
% 1  1  79.6779    76.9721    0.0307    0.3981
% 2  0  53.7836    54.0779    0.0532    0.3744
% 3  0  29.8613    35.1929    0.1681    0.2852
% 4  0  12.7375    20.2619    0.4290    0.1491
% 5  0  4.5017     9.1644     0.4073    0.0845
```

Diagnostic checks highlight that the null hypothesis of no autocorrelation of residuals cannot be rejected for each of the time series under analysis. Likewise, the normality test cannot be rejected. The reader interested in performing additional analysis aimed at detecting atypical observation may refer to Bellini (2016).

(Continued)

Example 2.3 VEC Analysis—cont'd

```
% 3. Misspecification Analysis
% Residuals
res=mles.r2.res;
resmean=mean(res);
resstdev=std(res);
resst=zeros(size(res,1),size(res,2));
for j=1:size(macv,2)
    resst(:,j)=(res(:,j)-resmean(1,j))/resstdev(1,j);
end
eU=resst';
alpha=0.05; %significance for the tests
p=2; % model order
h=1; %number of lags
% WHITENESS: portmanteau Ljung-Box test
[pval,Qh,critlo,crithi,crit1tail,CO,stringWhite,...
flagWhite]=test_whiteness(eU,p,h,alpha);
% Output
% non-rejection: signals are WHITE
% NONGAUSSIANITY of W residuals
[pGauss,ps,pk,lambdas,lambdak,crittresh,stringGauss,...
flagGauss]=test_gaussianity(resst',alpha);
%Output
% non-rejection: signals are GAUSSIAN
```

The appendix highlights how robust statistics tackles the issues related to anomalous units or cluster of observations. Specific emphasis is devoted to the forward search (FS) of Atkinson et al. (2004). Unlike other robust techniques, the FS aims to detect aberrant observations with suggestions for model enhancement by use of a graphical approach (Atkinson and Riani, 2000). The key advantage of the FS is that it is easy to interpret and implement. Indeed, it does not require any additional assumptions apart from those that characterize the model under analysis.

As indicated in Chapter 1, stress testing and risk integration rely on scenarios. Hence forecasting the dynamic of macroeconomic variables is a central topic in our time series analysis. In this regard, the next section focuses on the main theoretical assumptions to project VAR and VEC models.

2.3.2 Vector Autoregression and Vector Error-Correction Forecast

Let us start the forecast analysis from the following one-step-ahead predictor for a VAR model:

$$\mathbf{x}_{t+1|t} = \mathbf{c} + \boldsymbol{\Phi}_1 \mathbf{x}_{t|t} + \cdots + \boldsymbol{\Phi}_k \mathbf{x}_{t-k+1|t}. \tag{2.28}$$

A forecast for a longer horizon h relies on the following chain rule:

$$\mathbf{x}_{t+h|t} = \mathbf{c} + \mathbf{\Phi}_1 \mathbf{x}_{t-1+h|t} + \cdots + \mathbf{\Phi}_k \mathbf{x}_{t-k+h|t}. \tag{2.29}$$

The h-step forecast errors is

$$\mathbf{x}_{t+h} - \mathbf{x}_{t+h|t} = \sum_{r=0}^{h-1} \boldsymbol{\psi}_r \boldsymbol{\epsilon}_{t-r+h}, \tag{2.30}$$

where the matrices $\boldsymbol{\psi}_r$ are determined by recursive substitution

$$\boldsymbol{\psi}_r = \sum_{l=1}^{k-1} \boldsymbol{\psi}_{r-l} \mathbf{\Phi}_l, \tag{2.31}$$

with $\boldsymbol{\psi}_0 = \boldsymbol{I}_p$ and $\mathbf{\Phi}_l = \mathbf{0}$ for $l > k$. Let us consider forecasting $\mathbf{x}_{t+h}$ when the parameters of the VAR process are estimated by using multivariate least squares. The best linear predictor is as follows:

$$\hat{\mathbf{x}}_{t+h|t} = \hat{\mathbf{\Phi}}_1 \hat{\mathbf{x}}_{t-1+h|t} + \cdots + \hat{\mathbf{\Phi}}_k \hat{\mathbf{x}}_{t-k+h|t}, \tag{2.32}$$

where $\hat{\mathbf{\Phi}}_{(.)}$ is the estimated parameter matrix.

Since the multivariate forecast errors are asymptotically normally distributed, $\mathbf{x}_{t+h}$ can be simulated by the generation of multivariate normal random variables with zero mean and covariance matrix $\hat{\mathbf{\Sigma}}(h) = \sum_{r=0}^{h-1} \hat{\boldsymbol{\xi}}_r \hat{\mathbf{\Sigma}} \hat{\boldsymbol{\xi}}_r'$, with $\hat{\boldsymbol{\xi}}_r = \sum_{\iota=1}^{r} \hat{\boldsymbol{\xi}}_{r-\iota} \hat{\mathbf{\Sigma}}_\iota$. Lutkepohl (2005) gives an approximation to mean square error $MSE(\mathbf{x}_{t+h} - \hat{\mathbf{x}}_{t+h|t})$ that may be interpreted as finite sample correction of the above detailed covariance matrix. Asymptotic confidence intervals for h-step-ahead forecast may be obtained by use of the normal distribution assumption. Therefore a practical way to forecast time series is to use simulations as outlined in Example 2.4.

Example 2.4 Multivariate Time Series Simulation and Forecast

The VEC model estimated in Example 2.3 can be represented as follows:

$$
\underbrace{\begin{bmatrix} \widehat{\Delta x}_{1,t} \\ \vdots \\ \widehat{\Delta x}_{p,t} \end{bmatrix}}_{\widehat{\Delta x_t}} = \underbrace{\begin{bmatrix} \hat{\alpha}_{1,1} & \cdots & \hat{\alpha}_{1,r} \\ \vdots & \ddots & \vdots \\ \hat{\alpha}_{p,1} & \cdots & \hat{\alpha}_{p,r} \end{bmatrix}}_{\hat{\alpha}} \underbrace{\begin{bmatrix} \hat{\beta}_{1,1} & \cdots & \hat{\beta}_{p,1} \\ \vdots & \ddots & \vdots \\ \hat{\beta}_{1,r} & \cdots & \hat{\beta}_{p,r} \end{bmatrix}}_{\hat{\beta}'} \underbrace{\begin{bmatrix} x_{1,t-1} \\ \vdots \\ x_{p,t-1} \end{bmatrix}}_{x_{t-1}} + \underbrace{\begin{bmatrix} \hat{c}_1 \\ \vdots \\ \hat{c}_r \end{bmatrix}}_{\hat{c}}
$$

$$
+ \underbrace{\begin{bmatrix} \hat{\gamma}_{1,1} & \cdots & \hat{\gamma}_{1,p} \\ \vdots & \ddots & \vdots \\ \hat{\gamma}_{p,1} & \cdots & \hat{\gamma}_{p,p} \end{bmatrix}}_{\hat{r}_1} \underbrace{\begin{bmatrix} \Delta x_{1,t-1} \\ \vdots \\ \Delta x_{p,t-1} \end{bmatrix}}_{\Delta x_{t-1}} + \underbrace{\begin{bmatrix} e_{1,t} \\ \vdots \\ e_{p,t} \end{bmatrix}}_{e_t},
$$

(Continued)

Example 2.4 Multivariate Time Series Simulation and Forecast—cont'd

where Δx_t is the $p \times 1$ vector containing y, π, eq, er, r^{ST}, and r^{LT} first differences and $\hat{\alpha}$, $\hat{\beta}$, $\hat{c}$, and $\hat{\Gamma}_1$ contain model parameters. e_t is the error vector based on model estimates. The latter has a $p \times p$ covariance matrix $\hat{\Sigma}$ that is crucial to simulate innovations and forecast the model. In this regard, the first step is to compute $\hat{x}_{t+1|t}$ as detailed below:

- Generate a $p \times 1$ vector of errors $(e_{sim,t+1})$ from a distribution $N(0, \hat{\Sigma})$.
- Compute $\widehat{\Delta x}_{t+1}$ as follows:

$$\widehat{\Delta x}_{t+1|t} = \hat{\alpha}\left(\hat{\beta}' x_t + \hat{c}\right) + \hat{\Gamma}_1 \Delta x_t + e_{sim,t+1}.$$

- The final step is

$$\hat{x}_{t+1|t} = x_t + \widehat{\Delta x}_{t+1|t}.$$

The above steps apply to the specific example under analysis. The reader is invited to derive a more general rule to be used in any setting.

Once an easily derive the h-step-ahead forecast by recursively adding $\widehat{\Delta x}_{t+h|t}$ to $\hat{x}_{t-1+h|t}$.

The following MATLAB code shows how to simulate the VEC model estimated in Example 2.3.

```
% Simulation
nSim=1000; % Number of simulations
ini=2; % Initial step
len=24; % Number of simulation steps
nrow=ini-1+len; % Number of rows of sim vector
ncol=size(macv,2); % Number of variables to simulate
yt2=macv(end-1,:);   % Initializing values time: t-1
yt1=macv(end,:);     % Initializing values time: t
% CVARsim function creates a matrix "nrow x ncol x nSim"
% Each simulation block is added on the right hand side
[ySimt]=CVARsiml(nSim,nrow,ncol,lagCVAR,yt2,yt1,...
SIGMA,A,B,B1,c0);
```

Fig. 2.4 presents the simulated patterns for 24 quarters. The starting point is the last observation of the historical time series. No constraints are imposed on the simulations. However thresholds should be set to avoid misalignments with economic experience (e.g., highly negative interest rates).

Example 2.4 Multivariate Time Series Simulation and Forecast—cont'd

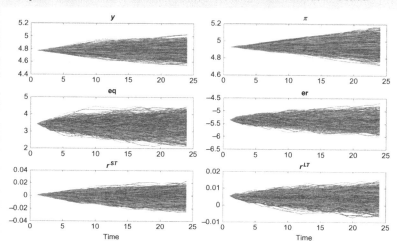

FIG. 2.4 Simulation of macroeconomic time series.

The last part of the forecast process based on Monte Carlo simulations relies on the following two steps. Firstly, compute the mean (or median) path. Secondly, estimate confidence bands as quantiles of the simulated distribution. Fig. 2.5 summarizes this process.

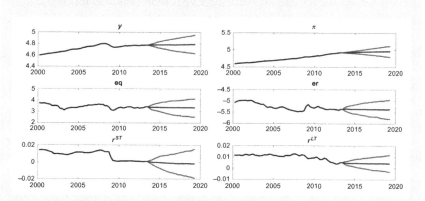

FIG. 2.5 Macroeconomic time series and forecast with corresponding confidence bands.

Exercise 2.4 shows how to fit a VEC model by use of the software R. Additionally, this is converted to a VAR model through the function *vec2var*.

When one is dealing with large sets of variables, extensive samples are required. Additionally, lagged values tend to produce high correlation and parameter estimates lose their robustness. Doan et al. (1984) proposed the use of Bayesian prior information (i.e., Minnesota prior). This approach has been widely used in practice. The reader may refer to LeSage (1999) for a useful discussion and detailed MATLAB implementation.

The next section introduces the impulse response analysis. This study is particularly useful in stress testing and reverse stress testing. Its main advantage is to capture the impact of a given macroeconomic variable shock on other correlated variables.

2.3.3   Impulse Response Analysis

As described in Eq. (2.3), an AR process can be represented as an infinite sum of past errors with decaying weights. In a similar way, a covariance stationary VAR process has a so-called Wold representation of the form

$$\mathbf{x}_t = \mu + \epsilon_t + \psi_1 \epsilon_{t-1} + \psi_2 \epsilon_{t-2} + \cdots, \tag{2.33}$$

where the $p \times p$ MA ψ_r is determined recursively with use of Eq. (2.31). One should be tempted to interpret the elements of ψ_r as the dynamic multiplier or impulse response. However, this is possible only if $\text{var}(\epsilon) = \Sigma$ is a diagonal matrix so that errors are uncorrelated. One way to obtain an uncorrelated error matrix is to estimate a triangular structural VAR represented as follows:

$$\mathbf{B}\mathbf{x}_t = \mathbf{c} + \Phi_1 \mathbf{x}_{t-1} + \cdots + \Phi_k \mathbf{x}_{t-k} + \eta_t, \tag{2.34}$$

where $\mathbf{B}$ is a lower triangular matrix with 1s along the diagonal. The uncorrelated orthogonal errors η_t are referred to as *structural errors*, and triangular structure imposes a recursive casual ordering. The ordering, for example, $\mathbf{x}_1 \rightarrow \mathbf{x}_2 \rightarrow \mathbf{x}_3$, imposes the restriction that $\mathbf{x}_1$ affects $\mathbf{x}_2$ and $\mathbf{x}_3$ but the latter variable may not affect $\mathbf{x}_1$. Hence the ordering depends on the context and its economic justification.

Given a recursive ordering, the Wold representation of $\mathbf{x}_t$ based on the orthogonal errors η_t is as follows:

$$\mathbf{x}_t = \mu + \Theta_0 \eta_t + \Theta_1 \eta_{t-1} + \Theta_2 \eta_{t-2} + \cdots, \tag{2.35}$$

where $\Theta_0 = \mathbf{B}^{-1}$ is a lower triangular matrix.

Exercise 2.4 shows how to implement the impulse response analysis in practice.

The global nature of economic interdependencies induces both researchers and practitioners to use GVAR models. The next section introduces the key concepts to develop this framework.

2.4   GLOBAL VECTOR AUTOREGRESSION MODELING

The multivariate time series models described in the previous sections are very useful to describe either a specific economy or a set of countries.

Pesaran et al. (2004) introduced GVAR modeling to embrace a more extended international environment. This approach is particularly useful for firms like Bank Alpha having a global footprint.

2.4.1   Introduction to the Global Vector Autoregression Model

Let us consider an economy with S countries indexed $s = 0, \ldots, S$. Country 0 is the reference country. A GVAR model is outlined starting from a typical VARX(2,2)[2] as follows:

$$
\begin{aligned}
\mathbf{x}_{s,t} =& \mathbf{c}_{s,0} + \mathbf{c}_{s,1}t + \mathbf{\Phi}_{s,1}\mathbf{x}_{s,t-1} + \mathbf{\Phi}_{s,2}\mathbf{x}_{s,t-2} \\
&+ \mathbf{\Lambda}_{s,0}\mathbf{x}_{s,t}^* + \mathbf{\Lambda}_{s,1}\mathbf{x}_{s,t-1}^* + \mathbf{\Lambda}_{s,2}\mathbf{x}_{s,t-2}^* + \mathbf{u}_{s,t},
\end{aligned}
\tag{2.36}
$$

where $\mathbf{x}_{s,t}$ is the $p_s \times 1$ vector of country-specific variables, $\mathbf{x}_{s,t}^*$ is the $p_s^* \times 1$ vector of foreign variables, $\mathbf{c}_{s,0}$ is a constant term, $\mathbf{c}_{s,1}$ is the trend coefficient, $\mathbf{\Phi}_{s,(\cdot)}$ and $\mathbf{\Lambda}_{s,(\cdot)}$ are the matrix coefficients for the domestic and foreign variables, and $\mathbf{u}_{s,t}$ represents the error component, which is assumed to be a serially uncorrelated and cross-sectionally weakly dependent process.

The foreign variable vectors have the following weighting structure:

$$
\mathbf{x}_{s,t}^* = \sum_{r=0}^{S} \mathbf{w}_{s,r}\mathbf{x}_{r,t}, \quad \mathbf{w}_{ss} = 0,
\tag{2.37}
$$

where $w_{s,r}$ are weights such that $\sum_{r=0}^{S} w_{s,r} = 1$. Weights can change over time. However, a fixed weight scheme facilitates the discussion. Foreign-specific variables are weighted averages of the corresponding domestic ones. Weights are set according to international trade flows.

The error correction form of the VARX(2,2) model outlined in Eq. (2.36) is represented as follows:

$$
\begin{aligned}
\Delta\mathbf{x}_{s,t} =& \mathbf{c}_{s,0} - \boldsymbol{\alpha}_s\boldsymbol{\beta}_s' \left[\boldsymbol{\zeta}_{s,t-1} - \boldsymbol{\gamma}_s(t-1)\right] \\
&+ \mathbf{\Lambda}_{s,0}\Delta\mathbf{x}_{s,t}^* + \mathbf{\Gamma}_s\Delta\boldsymbol{\zeta}_{s,t-1} + \mathbf{u}_{s,t},
\end{aligned}
\tag{2.38}
$$

where $\boldsymbol{\zeta}_{s,t} = \left(\mathbf{x}_{s,t}', \mathbf{x}_{s,t}^{*'}\right)'$, $\boldsymbol{\alpha}_s$ is a $p_s \times r_s$ matrix of rank r_s, and $\boldsymbol{\beta}_s$ is a $\left(p_s + p_s^*\right) \times r_s$ matrix or rank r_s.

For each country, models are estimated separately conditional on $\mathbf{x}_{s,t}^*$. The latter is treated as weakly exogenous with regard to the parameters in Eq. (2.38). Even though the estimation is done on a country basis, the GVAR model

---

2. VARX(2,2) stands for VAR with two lags on the domestic country and two lags on the foreign countries.

is solved for the international economy as a whole. Conditional on a given estimate of $\boldsymbol{\beta}_s$, the remaining parameters are estimated via ordinary least squares regressions as follows:

$$\Delta\mathbf{x}_{s,t} = \mathbf{c}_{s,0} + \delta_s ECM_{s,t-1} + \boldsymbol{\Lambda}_{s,0}\Delta\mathbf{x}^*_{s,t} + \boldsymbol{\Gamma}_s\Delta\mathbf{x}_{s,t-1} + \mathbf{u}_{s,t}, \qquad (2.39)$$

where $ECM_{s,t-1}$ are the error correction terms of the sth country model.

The next section describes how to estimate the GVAR model described above with the use of the MATLAB toolbox and the time series data available from https://sites.google.com/site/gvarmodelling/gvar-toolbox.

2.4.2 Global Vector Autoregression Analysis

The purpose of a stress test is to assess a bank's resilience under adverse conditions. A GVAR model may be used to project macroeconomic variables and enrich a given scenario.

The aim of this section is to estimate a GVAR model to use for Bank Alpha's stress testing and risk integration analysis. The analysis relies on 33 countries representing the major economies of the world. Table 2.4 summarizes the sample and highlights that eight countries are treated as one area (i.e., euro area).

A transformation of the following variables feeds each individual VARX(p_s, q_s) model: nominal GDP of country s at time t in local currency $(GDP_{s,t})$: CPI for country s at time t $(CPI_{s,t})$; exchange rate of the currency of country s at time t in US dollars $(ER_{s,t})$; nominal equity price index $(EQ_{s,t})$; nominal short-term interest rate per annum, in percent $(R^{ST}_{s,t})$; nominal long-

TABLE 2.4 Global Vector Autoregression Countries

All Countries			Euro Area
Argentina	India	Philippines	Austria
Australia	Indonesia	South Africa	Belgium
Austria	Italy	Saudi Arabia	Finland
Belgium	Japan	Singapore	France
Brazil	Korea	Spain	Germany
Canada	Malaysia	Sweden	Italy
China	Mexico	Switzerland	Netherlands
Chile	Netherlands	Thailand	Spain
Finland	Norway	Thailand	
France	New Zealand	United Kingdom	
Germany	Peru	United States	

TABLE 2.5 Global Vector Autoregression Core Variables

Descriptions	Symbols and Analytical Formulas
Real output	$y_{s,t} = \ln(GDP_{s,t}/CPI_{s,t})$
In CPI	$\pi_{s,t} = \ln(CPI_{s,t})$
Real exchange rate	$er_{s,t} = \ln(ER_{s,t}/CPI_{s,t})$
Real equity price	$eq_{s,t} = \ln(EQ_{s,t}/CPI_{s,t})$
Short-term interest rate	$r_{s,t}^{ST} = 0.25\ln(1 + R_{s,t}^{ST}/100)$
Long-term interest rate	$r_{s,t}^{LT} = 0.25\ln(1 + R_{s,t}^{LT}/100)$
Oil price	$p_{oil,t} = \ln(P_{oil,t})$

term interest rate per annum, in percent ($R_{s,t}^{LT}$); oil price in US dollars ($p_{oil,t}$). The quarterly time series cover the period from 1979 to 2013. Table 2.5 outlines variable transformations and their corresponding abbreviations (symbols).

The United States is indexed as country 0, and the exchange rate of the United States $er_{0,t}$ is taken to be 1. The endogenous and foreign variables can be summarized as follows:

- **Variables for all countries except the United States.** According to data availability, for all countries other than the United States the endogenous variables are $y_{s,t}$, $\pi_{s,t}$, $er_{s,t}$, $eq_{s,t}$, $r_{s,t}^{ST}$, and $r_{s,t}^{LT}$. The foreign variables are $y_{s,t}^*$, $\pi_{s,t}^*$, $eq_{s,t}^*$, $r_{s,t}^{ST,*}$, $r_{s,t}^{LT,*}$, and $p_{oil,t}$.
- **Variables for the United States.** For the United States the endogenous variables can be summarized as follows: $y_{US,t}$, $\pi_{US,t}$, $eq_{US,t}$, $r_{US,t}^{ST}$, $r_{US,t}^{LT}$, and $p_{oil,t}$. The exogenous variables are $y_{US,t}^*$, $\pi_{US,t}^*$, and $er_{US,t}^*$.

The weights $\mathbf{w}_{s,r}$ are trade weights between country s and country r computed as the average of monthly trades of a country during the period from 2009 to 2011.

As anticipated earlier, the estimation is done on a country-specific basis and then the model is solved as a whole. A useful starting point to show this process is to consider the VARX(2,2) model described in Eq. (2.36). $\boldsymbol{\zeta}_{s,t} = (\mathbf{x}_{s,t}', \mathbf{x}_{s,t}^{*'})'$ is used to rewrite the equation as follows:

$$\mathbf{A}_{s,0}\boldsymbol{\zeta}_{s,t} = \mathbf{c}_{s,0} + \mathbf{c}_{s,1}t + \mathbf{A}_{s,1}\boldsymbol{\zeta}_{s,t-1} + \mathbf{A}_{s,2}\boldsymbol{\zeta}_{s,t-2} + \mathbf{u}_{s,t}, \qquad (2.40)$$

where $\mathbf{A}_{s,0} = (\mathbf{I}_{p,s}, -\boldsymbol{\Lambda}_{s,0})$, $\mathbf{A}_{s,1} = (\boldsymbol{\Phi}_{s,1}, \boldsymbol{\Lambda}_{s,1})$, and $\mathbf{A}_{s,2} = (\boldsymbol{\Phi}_{s,2}, \boldsymbol{\Lambda}_{s,2})$.

The link matrix $\mathbf{W}_s$, defined by the country-specific trade weights, can be used to obtain the following identity:

$$\boldsymbol{\zeta}_{s,t} = \mathbf{W}_s\mathbf{x}_t, \qquad (2.41)$$

where $\mathbf{x}_t = (\mathbf{x}'_{0,t}, \mathbf{x}'_{1,t}, \ldots)'$ is the matrix collecting all the endogenous variables of the system. Eq. (2.40) is rewritten as

$$\mathbf{A}_{s,0}\mathbf{W}_s\mathbf{x}_t = \mathbf{c}_{s,0} + \mathbf{c}_{s,1}t + \mathbf{A}_{s,1}\mathbf{W}_s\mathbf{x}_{t-1} + \mathbf{A}_{s,2}\mathbf{W}_s\mathbf{x}_{t-2} + \mathbf{u}_{s,t}, \qquad (2.42)$$

and these individual models are then used as listed below:

$$\mathbf{G}_0\mathbf{x}_t = \mathbf{c}_0 + \mathbf{G}_1\mathbf{x}_{t-1} + \mathbf{G}_2\mathbf{x}_{t-2} + \mathbf{u}_{s,t}, \qquad (2.43)$$

where $\mathbf{G}_0 = [(\mathbf{A}_{0,0}\mathbf{W}_0), (\mathbf{A}_{1,0}\mathbf{W}_1), \ldots]'$, $\mathbf{G}_1 = [(\mathbf{A}_{0,1}\mathbf{W}_0), (\mathbf{A}_{1,1}\mathbf{W}_1), \ldots]'$, $\mathbf{G}_2 = [(\mathbf{A}_{0,2}\mathbf{W}_0), (\mathbf{A}_{1,2}\mathbf{W}_1), \ldots]'$, $\mathbf{c}_0 = [\mathbf{c}_{0,0}, \mathbf{c}_{1,0}, \ldots]'$, $\mathbf{u}_0 = [\mathbf{u}_{0,t}, \mathbf{u}_{1,t}, \ldots]'$.

The matrix $\mathbf{G}_0$ is a nonsingular matrix depending on trade weights and parameter estimates. Therefore it is allowed to premultiply Eq. (2.43) by $\mathbf{G}_0^{-1}$ to obtain the following equation:

$$\mathbf{x}_t = \mathbf{b}_0 + \mathbf{b}_1 t + \mathbf{F}_1\mathbf{x}_{t-1} + \mathbf{F}_2\mathbf{x}_{t-2} + \boldsymbol{\epsilon}_t, \qquad (2.44)$$

where $\mathbf{F}_1 = \mathbf{G}_0^{-1}\mathbf{G}_1$, $\mathbf{F}_2 = \mathbf{G}_0^{-1}\mathbf{G}_2$, $\mathbf{b}_0 = \mathbf{G}_0^{-1}\mathbf{c}_0$, $\mathbf{b}_1 = \mathbf{G}_0^{-1}\mathbf{c}_1$, and $\boldsymbol{\epsilon}_t = \mathbf{G}_0^{-1}\mathbf{u}_t$.

The estimation of all the above equations is performed through the MATLAB GVAR toolbox. This software allows the user to select the countries to be included in the model and the corresponding variables. Additionally, a series of parameterizations are available. Our study is conducted on the countries described in Table 2.4 and the variables are indicated in Table 2.5 . The key results of the estimation process are summarized as follows:

- **Stationarity test.** Unit root tests such as the Dickey-Fuller and weighted symmetric ADF tests suggest that for most of the variables the null hypothesis of nonstationarity cannot be rejected.
- **Estimation-selection.** The choice of domestic and foreign lag orders (k_s, k_s^*) of the individual VARX* models relies on the AIC with $k_{s,max} = 2, k_{s,max}^* = 2$. For most countries a VARX*$(2, 1)$ specification is satisfactory. VEC country-specific models are estimated on the basis of reduced-rank regression. The rank of the cointegrating space is estimated by use of the Johansen trace statistic as detailed by Pesaran et al. (2004). Two cointegrated relations are detected for the key Bank Alpha business areas: the United States, China, the euro area, and Japan. For the United kingdom, three cointegrating relations are shown by the model.
- **Diagnostic.** Serial correlation analyses based on F-test statistics show that the null hypothesis of serial correlation at the 5% significance level is rejected for most of the variables under analysis.
- **Weak exogeneity.** The modeling analysis continues by checking the assumption that the country-specific foreign variables are weakly exogenous $I(1)$. This assumption is verified for the vast majority of countries, and Table 2.6 shows the F-test for Bank Alpha key business countries.

All parameter estimates with their statistics are stacked by the GVAR toolbox in an Excel spreadsheet. One can use these estimates to perform

TABLE 2.6 Weak Exogeneity F Statistics. Countries and Regions Where Bank Alpha has its Core Business

Country/Region	F-test	$F_{0.05}$	y^*	π^*	eq^*	er^*	$r_{s,t}^{ST^*}$	$r_{s,t}^{LT^*}$	p_{oil}
United States	$F(2,119)$	3.07	0.31	2.84		0.47			1.32
China	$F(2,117)$	3.07	0.56	0.58	0.22		1.14	1.67	1.17
Euro area	$F(2,115)$	3.08	1.15	4.29	0.38		3.25^a	0.20	1.80
Japan	$F(2,115)$	3.08	1.67	2.16	0.71		0.43	1.09	0.57
United Kingdom	$F(3,114)$	2.68	2.70^a	0.35	0.74		0.28	0.23	0.52

^{a} Statistical significance at the 5% level.

additional analysis. Moreover, the toolbox allows one to forecast the estimated model, as outlined in the next section.

2.4.3 Global Vector Autoregression Forecast

A useful starting point to forecast a GVAR model is Eq. (2.44) expressed in companion form as follows:

$$
\begin{bmatrix} \mathbf{x}_t \\ \mathbf{x}_{t-1} \end{bmatrix} = \begin{bmatrix} \mathbf{F}_1 & \mathbf{F}_2 \\ \mathbf{I}_p & \mathbf{0} \end{bmatrix} \begin{bmatrix} \mathbf{x}_{t-1} \\ \mathbf{x}_{t-2} \end{bmatrix} + \begin{bmatrix} \mathbf{b}_0 + \mathbf{b}_1 t \\ \mathbf{0} \end{bmatrix} + \begin{bmatrix} \boldsymbol{\epsilon}_t \\ \mathbf{0} \end{bmatrix}, \quad (2.45)
$$

or

$$
\mathbf{x}_t = \mathbf{F}\mathbf{x}_{t-1} + \mathbf{D}_t + \mathbf{V}_t. \quad (2.46)
$$

Hence

$$
\mathbf{x}_{t+h|t} = \mathbf{F}_h \mathbf{x}_t + \sum_{l=0}^{h-1} \mathbf{F}_l \mathbf{D}_{t+h-l|t} + \sum_{l=0}^{h-1} \mathbf{F}_l \mathbf{V}_{t+h-l|t}. \quad (2.47)
$$

Therefore, conditional on the initial value $\mathbf{x}_t$, the point forecast is as follows:

$$
\boldsymbol{\mu}_{h|t} = \mathbf{E}_1 \mathbf{F}_h \mathbf{x}_t + \sum_{l=0}^{h-1} \mathbf{E}_1 \mathbf{F}_l \mathbf{D}_{t+h-l|t}, \quad (2.48)
$$

where $\mathbf{E}_1 = (\mathbf{I}_p, \mathbf{0}_{p \times p})$.

According to the above procedure, the forecast function of the GVAR toolbox allows one to obtain the projection for all countries and variables under analysis. Fig. 2.6 shows the dynamics of the GDP for the key countries where Bank Alpha operates. A simulation process allows one to obtain confidence intervals within which the expected projection is likely to lie with a given probability.

An extension of the impulse response analysis to a more generalized framework to be used for stress testing purposes is detailed in the next section.

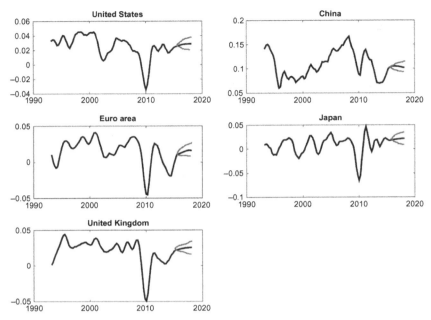

FIG. 2.6 GDP growth historical time series and global vector autoregression forecast.

2.4.4 Generalized Impulse Response Analysis

The traditional impulse response analysis as described in Section 2.3.3 requires orthogonalization of shocks and is not invariant to the ordering of the variables in the VAR. Pesaran and Shin (1998) proposed the following generalized impulse response function (GIRF) that is not subject to the above limitations:

$$GIRF(\mathbf{x}_t | u_{s,l,t}, h) = \mathbb{E}(\mathbf{x}_{t+h} | u_{s,l,t} = \sqrt{\sigma_{ss,ll}}, I_{t-1}) - \mathbb{E}(\mathbf{x}_{t+h} | I_{t-1}), \qquad (2.49)$$

where I_{t-1} is the information at time $t-1$ and $\sigma_{ss,ll}$ is the diagonal element of the covariance matrix Σ_u. It corresponds to the lth equation of the sth country. h is the forecasting time horizon. According to this definition, a one standard error shock at time t on variable r, has the following impact on variable l at time $t+h$:

$$GIRF(\mathbf{x}_t | u_{s,l,t}, h) = \frac{\mathbf{e}_r' \mathbf{A}_h \mathbf{G}_0^{-1} \Sigma_u \mathbf{e}_l}{\sqrt{\mathbf{e}_l' \Sigma_u \mathbf{e}_l}} \qquad (2.50)$$

In what follows, an illustrative example of the GIRF analysis based on a US GDP three standard deviations negative shock is provided. As per Bank Alpha's core business, the focus is on the United States, China, the euro area, Japan, and the United Kingdom. Fig. 2.7 highlights that a GDP negative shock has an immediate impact on all the economies under analysis. The Chinese economy seems to be able to return promptly to its original growth level. The lower GIRF

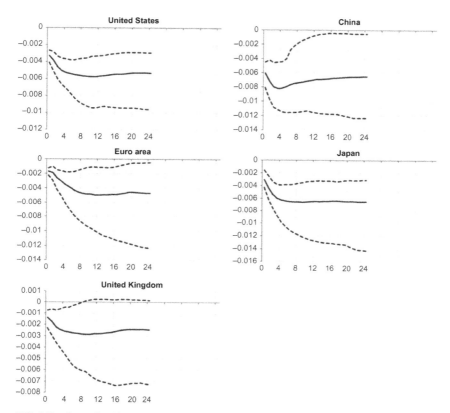

FIG. 2.7 Generalized impulse response function analysis based on a negative shock to US GDP growth rate. Median, lower, and upper bootstrap bounds.

thresholds of the euro area and Japan result in the most persistent damage from the adverse scenario. The UK economy recovers after the shock. However, wide confidence bounds capture a high uncertainty around the central path.

The next section outlines the key features of a regulatory scenario used throughout the book. It also proposes statistical tools for its enrichment (i.e., conditional forecast).

2.5   STRESS TESTING SCENARIO

A stress scenario is not a forecast of macroeconomic and financial conditions. On the contrary, it is a coherent set of conditions designed specifically to assess the resilience of banks when facing a deterioration in global economic conditions.

The following section outlines the key components of a scenario design. The reader will also learn about a statistical framework to enrich a scenario, based on a conditional forecast method.

2.5.1 Scenario Design

Mindful of a stress testing purpose, the narrative accompanying the exercise plays a key role. As an example, Business Case 2.1 summarizes the narrative for the Bank of England 2015 stress test.

Business Case 2.1   Narrative for the Bank of England 2015 Stress Test

A rapid deterioration of market sentiment characterizes the scenario. The risk appetite diminishes and a general derisk attitude spreads among market participants. Capital flows to high-quality US assets. The dollar appreciates against a wide range of currencies. Liquidity in some markets becomes seriously impaired, and credit risk premiums rise consistently. The price of commodities falls by causing downward pressure on global inflation. Policy support in China helps in rebalancing the economy toward consumption. However, this does not occur immediately. Property prices fall sharply. In turn, investment in residential property and associated industries contracts. Growth slows materially and the renminbi depreciates against the dollar. In the euro area a disinflationary pressure is caused by weaker domestic demand. Consumption and investment decisions are delayed. This amplifies the downturn in activity. Unemployment increases materially throughout the euro area. A significant increase in credit risk premiums characterizes the most highly indebted sovereign states, households, and firms. These global shocks have adverse implications for activity in a number of emerging economies. This is particularly significant for China's major trading partners. These countries also experience higher risk premiums on foreign borrowing, which triggers a sudden stop of capital inflows and a sharp contraction of domestic credit and demand. The appreciation of the dollar affects all business denominated in that currency. Global downturn impacts the UK economy. Output growth turns negative as export demand falls sharply. The household and corporate savings rates increase because of precautionary behavior and the higher cost of credit as banks face higher funding costs. Falls in consumption, investment, and property prices characterize the scenario. A sharp rise in risk premiums affects private sector borrowers. Policymakers observe these developments as a series of unexpected shocks. Additional monetary policy stimulus takes the form of lowering the yield curve over the course of the stress scenario.

In line with the above, regulators supply macroeconomic scenarios by distinguishing between baseline and adverse (in some cases even severe adverse). They provide projections for a restricted set of variables (e.g., GDP, unemployment rate, and so on). However, some of the variables required by banks to stress their own risk parameters may not be included in the regulatory set. Conditional forecasting models provide a support to enrich regulatory scenarios. Indeed, they allow one to include variables other than those given by regulators by relying on a coherent projection framework, as detailed in the next section.

2.5.2 Conditional Forecasting

Data availability is often an issue when one is dealing with time series analysis. The approach used for incomplete data is applicable to conditional forecasting as detailed below.

Let us consider the case where for some variables data are available up to t, while for others they are available up to $t + 2$. This may be caused by a delay in publishing statistics, misalignments, and so on. Thus one can make a forecast for the missing observations using a partially complete dataset (Robertson and Tallman, 1999).

A useful starting point is the VAR process with two variables and one lag studied in Example 2.2:

$$\begin{bmatrix} x_{1,t} \\ x_{2,t} \end{bmatrix} = \begin{bmatrix} 0.2 \\ 0.2 \end{bmatrix} + \begin{bmatrix} 0.3 & 0.2 \\ 0.2 & 0.3 \end{bmatrix} \begin{bmatrix} x_{1,t-1} \\ x_{2,t-1} \end{bmatrix} + \begin{bmatrix} \epsilon_{1,t} \\ \epsilon_{2,t} \end{bmatrix}, \tag{2.51}$$

where the error covariance matrix Σ can be factorized as

$$\Sigma = \begin{bmatrix} 1.0 & 0.8 \\ 0.8 & 1.0 \end{bmatrix} = \begin{bmatrix} 1.0 & 0.0 \\ 0.8 & 0.6 \end{bmatrix} \begin{bmatrix} 1.0 & 0.8 \\ 0.0 & 0.6 \end{bmatrix}.$$

Given the above covariance structure, one can rewrite Eq. (2.51) as

$$\begin{bmatrix} x_{1,t} \\ x_{2,t} \end{bmatrix} = \begin{bmatrix} 0.2 \\ 0.2 \end{bmatrix} + \begin{bmatrix} 0.3 & 0.2 \\ 0.2 & 0.3 \end{bmatrix} \begin{bmatrix} x_{1,t-1} \\ x_{2,t-1} \end{bmatrix} + \begin{bmatrix} 1.0 & 0.0 \\ 0.8 & 0.6 \end{bmatrix} \begin{bmatrix} \varepsilon_{1,t} \\ \varepsilon_{2,t} \end{bmatrix}, \tag{2.52}$$

where the vector $(\varepsilon_{1,t}, \varepsilon_{2,t})'$ contains normal errors with mean zero and covariance equal to an identity matrix.

Now let us suppose that information on $x_{1,t+1}$ and $x_{1,t+2}$ is available, while $x_{2,t}$ is the most recent data for the other variable. Thus the $t+1$ and $t+2$ equations may be written as

$$\begin{bmatrix} x_{1,t+1} \\ x_{2,t+1} \end{bmatrix} = \begin{bmatrix} 0.2 \\ 0.2 \end{bmatrix} + \begin{bmatrix} 0.3 & 0.2 \\ 0.2 & 0.3 \end{bmatrix} \begin{bmatrix} x_{1,t} \\ x_{2,t} \end{bmatrix} + \begin{bmatrix} 1.0 & 0.0 \\ 0.8 & 0.6 \end{bmatrix} \begin{bmatrix} \varepsilon_{1,t+1} \\ \varepsilon_{2,t+1} \end{bmatrix} \tag{2.53}$$

and

$$\begin{bmatrix} x_{1,t+2} \\ x_{2,t+2} \end{bmatrix} = \begin{bmatrix} 0.2 \\ 0.2 \end{bmatrix} + \begin{bmatrix} 0.3 & 0.2 \\ 0.2 & 0.3 \end{bmatrix} \begin{bmatrix} x_{1,t} \\ x_{2,t} \end{bmatrix} + \begin{bmatrix} 1.0 & 0.0 \\ 0.8 & 0.6 \end{bmatrix} \begin{bmatrix} \varepsilon_{1,t+1} \\ \varepsilon_{2,t+1} \end{bmatrix} \tag{2.54}$$

$$+ \begin{bmatrix} 1.0 & 0.0 \\ 0.8 & 0.6 \end{bmatrix} \begin{bmatrix} \varepsilon_{1,t+2} \\ \varepsilon_{2,t+2} \end{bmatrix}.$$

With data up to t, the best guess about the future errors is zero. Hence the forecast of $t + 1$ and $t + 2$ would correspond to values of the two variables at time t. Nonetheless, the forecast may benefit from all available information (i.e., $(x_{1,t+1}, x_{1,t+2})'$).

According to Robertson and Tallman (1999), the difference between the forecast of x_1 and its actual values is stacked in a 2×1 vector as follows:

$$\mathbf{r} = \left[\begin{array}{c} x_{1,t+1} - x_{1,t} \\ x_{1,t+2} - x_{1,t} \end{array} \right]. \tag{2.55}$$

It is worth highlighting that the first row of $\mathbf{r}$ corresponds to ε_{t+1}, while the second row is $(\varepsilon_{t+1} + \varepsilon_{t+2})$. Thus there is only one value of x_{t+1} and x_{t+2} that satisfies each constraint.

The process to outline how to align with each constrain starts from the definition of the following vector of errors:

$$\boldsymbol{\varepsilon} = \left[\begin{array}{c} \varepsilon_{1,t+1} \\ \varepsilon_{2,t+1} \\ \varepsilon_{1,t+2} \\ \varepsilon_{2,t+2} \end{array} \right]. \tag{2.56}$$

Additionally, let us consider the following matrix:

$$\mathbf{R} = \left[\begin{array}{cccc} 1 & 0 & 0 & 0 \\ 1 & 0 & 1 & 0 \end{array} \right], \tag{2.57}$$

where each row of $\mathbf{R}$ represents the relationship between $\mathbf{r}$ and $\boldsymbol{\varepsilon}$. Thus a system of equations of the form $\mathbf{r} = \mathbf{R}\boldsymbol{\varepsilon}$, to be solved for $\boldsymbol{\varepsilon}$, gives infinitely many possible solutions to the constrained forecast. Doan et al. (1984) highlighted that a unique error vector that satisfies the constraints and minimizes the sum of squares $\boldsymbol{\varepsilon}'\boldsymbol{\varepsilon}$ is

$$\hat{\boldsymbol{\varepsilon}} = \left[\begin{array}{c} \hat{\varepsilon}_{1,t+1} \\ 0 \\ \hat{\varepsilon}_{1,t+2} \\ 0 \end{array} \right], \tag{2.58}$$

where $\hat{\varepsilon}_{1,t+1} = x_{1,t+1} - x_{1,t}$ and $\hat{\varepsilon}_{1,t+2} = x_{1,t+2} - x_{1,t} - \hat{\varepsilon}_{1,t+1}$. As a consequence, it is convenient to substitute the elements of $\hat{\boldsymbol{e}}$ in the VAR model. Hence the known values are considered instead of $x_{1,t}$ as the forecast of $x_{1,t+h}$ ($h = 1, 2$). In line with the example above (i.e., Eqs. (2.53) and (2.54)) the following values apply to forecast $x_{2,t+h}$. More specifically $x_{2,t+1} = 0.2 + 0.2x_{1,t} + 0.3x_{2,t} + 0.8\hat{\varepsilon}_{1,t+1}$ and $x_{2,t+2} = 0.2 + 0.2x_{1,t} + 0.3x_{2,t} + 0.8(\hat{\varepsilon}_{1,t+1} + \hat{\varepsilon}_{1,t+2})$. This procedure is used recursively to constrain VAR models. This process allows the analyst to enrich the stress testing scenario and include all variables required by the bank.

More in general, let us consider a model where k forecasts are constrained at values $a_1, \ldots, a_k$. The value of the required errors may be computed by use of the following sorted equations, where constraints are listed on the top k equations as follows:

$$
\begin{bmatrix} a_1 \\ \vdots \\ a_k \\ x_t \\ \vdots \\ x_m \end{bmatrix} = f(\hat{\Theta}, \mathbf{x}_t) + \begin{bmatrix} r_1 \\ \vdots \\ r_k \\ \hat{\varepsilon}_t \\ \vdots \\ \hat{\varepsilon}_m \end{bmatrix}, \tag{2.59}
$$

where $\mathbf{r} = (r_1, \ldots, r_k)'$, the vector of error is defined as $\boldsymbol{\varepsilon} = (\varepsilon_1, \ldots, \varepsilon_m)'$, and the $\mathbf{R}$ matrix is

$$
\underset{(k \times m)}{\mathbf{R}} = \begin{bmatrix} 1 & 0 & \ldots & 0 & 0 & 0 & \ldots & 0 \\ 0 & 1 & \ldots & 0 & 0 & 0 & \ldots & 0 \\ \vdots & \vdots & \ddots & \vdots & \vdots & \vdots & \vdots & 0 \\ 0 & 0 & \ldots & 1 & 0 & 0 & \ldots & 0 \end{bmatrix}, \tag{2.60}
$$

$$\underbrace{\hphantom{XXXXXXXXXXXXXX}}_{m}$$

and the solution proposed by Robertson and Tallman (1999) is

$$
\hat{\varepsilon} = \mathbf{R}'(\mathbf{R}\mathbf{R}')^{-1}\mathbf{r} \tag{2.61}
$$

$$
= \begin{bmatrix} 1 & 0 & \ldots & 0 \\ 0 & 1 & \ldots & 0 \\ \vdots & \vdots & \ddots & \vdots \\ 0 & 0 & \ldots & 1 \\ 0 & 0 & \ldots & 0 \\ \vdots & \vdots & \ddots & \vdots \\ 0 & 0 & \ldots & 0 \end{bmatrix} \begin{bmatrix} 1 & 0 & \ldots & 0 \\ 0 & 1 & \ldots & 0 \\ \vdots & \vdots & \ddots & \vdots \\ 0 & 0 & \ldots & 1 \end{bmatrix} \begin{bmatrix} a_1 \\ \vdots \\ a_k \end{bmatrix}
$$

$$
= \begin{bmatrix} a_1 & \ldots & a_k & 0 & \ldots & 0 \end{bmatrix}'.
$$

As a primary goal of the stress testing, the next section summarizes the macroeconomic scenario on which to perform Bank Alpha stress testing.

2.5.3 Bank Alpha's Stress Testing Scenario

The idea behind Bank Alpha's stress testing scenario is to reflect the following main systemic risks:

- A deep crisis in the United States causes a contraction of all major economies. A reduction in GDP causes an increase in unemployment. Inflation and interest rates remain stable at a low level.
- The credit quality deteriorates as a consequence of the global crisis.
- Stalling policy reforms jeopardize confidence in the sustainability of public finances and accommodating monetary policies.

Table 2.7 summarizes the scenario to use for Bank Alpha's stress testing exercise. The focus is on the countries where Bank Alpha conducts its business. A negative shock originating from the United States threatens all economies. After a substantial GDP reduction in the first quarter, China and developing Asia promptly recover. The euro area and the United Kingdom are deeply affected throughout the first year. A positive response characterizes these economies from the fifth quarter onward. Japan is the only country suffering for a longer period after the initial negative shock.

The next section outlines the transmission mechanism connecting macroeconomic and bank-specific variables.

2.5.4 Macroeconomic Modeling and Satellite Frameworks

The preceding macroeconomic modeling is the first step of a process to assess bank resilience against adverse economic conditions. Fig. 2.8 shows the link between this chapter and the following ones. A distinction is made between direct and indirect system model feeding. More precisely, a direct relationship links a macroeconomic scenario and interest rates, market risk parameters, and so on. In contrast, a chain rule mechanism applies to all other ingredients of the stress test engine as listed below:

- **Direct feeding.** A macroeconomic scenario directly feeds the following models:

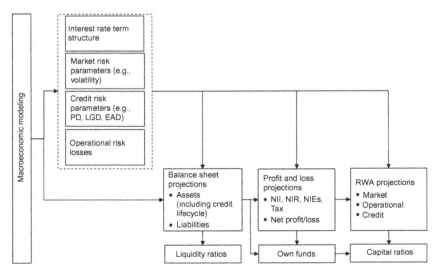

FIG. 2.8 Macroeconomic modeling relationship with satellite models. *EAD*, Exposure at default; *LGD*, loss given default; *NIEs*, noninterest expenses; *NII*, net interest income; *NIR*, noninterest revenue; *PD*, probability of default; *RWA*, risk-weighted asset.

TABLE 2.7 Stress Testing Adverse Scenario

Time	US Real GDP Growth (%)	US Unemployment Rate (%)	US Inflation Rate (%)	US Short-Term Interest Rate (%)	US Long Term Interest Rate (%)	China and DA Real GDP Growth (%)	China and DA Inflation Rate (%)	China and DA USD ER[a]	Euro Area Real GDP Growth (%)	Euro Area Inflation Rate (%)	Euro Area USD ER	Japan Real GDP Growth (%)	Japan Inflation Rate (%)	Japan USD ER	UK Real GDP Growth (%)	UK Inflation Rate (%)	UK USD ER
Q1	-4.27	5.60	2.10	0.10	1.00	0.64	2.96	92.82	-4.55	0.49	1.05	-7.42	-1.40	96.14	-2.94	-0.28	1.50
Q2	-2.73	6.16	1.19	0.10	1.20	3.28	0.08	92.63	-2.52	-0.49	1.05	-5.95	-2.31	96.62	-2.24	-0.91	1.49
Q3	-2.24	6.65	0.91	0.10	1.30	4.64	-0.88	92.34	-1.05	-0.77	1.04	-4.48	-2.31	97.47	-1.40	-0.91	1.49
Q4	-1.05	6.93	0.77	0.10	1.50	5.28	-0.96	91.87	-0.07	-0.77	1.03	-3.08	-1.89	98.23	-0.56	-0.63	1.48
Q5	0.84	7.00	1.12	0.10	1.50	5.44	0.32	89.97	0.70	-0.21	1.03	-1.75	-1.05	98.23	0.21	–	1.48
Q6	0.84	7.07	1.33	0.10	1.60	5.28	1.04	87.97	1.19	0.14	1.04	-0.70	-0.49	97.95	0.91	0.42	1.48
Q7	2.10	7.00	1.40	0.10	1.80	5.20	1.44	86.17	1.47	0.35	1.05	0.14	-0.14	97.76	1.47	0.70	1.48
Q8	2.10	6.93	1.33	0.10	1.90	5.12	1.44	84.74	1.54	0.35	1.06	0.77	-0.07	97.76	1.82	0.84	1.48
Q9	2.73	6.79	1.33	0.10	2.00	5.04	1.52	83.98	1.54	0.42	1.07	1.19	–	97.85	2.10	0.98	1.47
Q10	2.73	6.65	1.19	0.10	2.10	5.04	1.36	83.51	1.54	0.35	1.08	1.47	–	97.95	2.17	0.98	1.47
Q11	2.73	6.51	1.12	0.10	2.20	5.12	1.36	83.13	1.40	0.35	1.09	1.61	0.07	97.95	2.24	0.98	1.47
Q12	2.73	6.37	1.12	0.10	2.30	5.12	1.52	82.84	1.33	0.42	1.09	1.68	0.28	97.76	2.17	1.05	1.47

[a] The exchange rate (ER) is an index.
DA, Developing Asian countries.

- Interest rate term structure. Chapter 3 outlines this model with specific reference to margin at risk, full fair value, and market value at risk.
- Market risk parameters (e.g., volatility, correlation). These parameters are investigated in Chapter 3 with regard to the value at risk.
- Credit risk parameters (e.g., probability of default, loss given default, exposure at default). Chapter 4 explores the connection between macroeconomic variables and credit risk parameters.
- Operational risk. The analysis of the relationship between operational losses and macroeconomic scenarios is covered in Chapter 5.
- Balance sheet projections. Bank strategy and macroeconomic dynamics drive balance sheet projections. Chapter 5 studies this relationship by using all the risk sources described above.
- **Indirect feeding**. Profit and loss projections, risk-weighted asset evolution, and the derived capital and liquidity ratios are the final ring of the stress test chain (see the boxes on the right-hand side of Fig. 2.8). Chapter 6 collects all stress testing model inputs. Chapters 7 and 8 introduce some additional modeling techniques to perform risk integration and reverse stress testing.

2.6 SUMMARY

The chapter introduced the basic concepts around univariate time series. AR, MA, and ARMA models were analyzed. The Box-Jenkins method proposed a practical approach to fit a model to real time series. The study of UK inflation showed how to specify, estimate, select, and perform diagnostic tests on real data with use of the MATLAB Econometric Toolbox.

The VAR model constituted a starting point for macroeconomic time series analysis. The distinction between integration of order zero and order greater than zero paved the way to the VEC modeling. A practical exemplification showed how to use the latter by fitting UK macroeconomic time series.

Model simulation, forecasting, and impulse-response analysis completed this introductory multivariate time series discussion.

Given the role of international interdependencies, the last part of the chapter focused on the GVAR model. The estimation performed on a short list of key domestic and foreign variables embraced a set of 33 countries. Consequently, a conditional forecast study served the scope of enriching a given regulatory path. A scenario to use for Bank Alpha's stress testing exercise concluded the last section together with a scheme describing the satellite models that will characterize the next chapters.

The exercises at the end of this chapter are a meaningful tool for readers to familiarize themselves with the techniques outlined above as well as MATLAB and R statistical software.

SUGGESTIONS FOR FURTHER READING

Time series analysis is described in classic books, for example, that of Hamilton (1994). Johansen (1996) and Juselius (2006) focus on cointegration by highlighting both theory and practical exemplifications. Additionally, Lutkepohl (2005) proposes a very interesting analysis of VAR and VEC models by means of JMulti (i.e., a software application explicitly devoted to VAR, VEC, and other time series analysis). Following a more marked software perspective, Pfaff (2008) shows how to analyze integrated and cointegrated time series with R.

Following Pesaran et al. (2004), the high number of recent articles on GVAR shows growing interest in this subject. Among these articles, it is worth mentioning Dees et al. (2007) and Castren et al. (2010). Additionally, a very useful review of theoretical and empirical GVAR modeling can be found in Di Mauro and Pesaran (2013). From a software standpoint, a MATLAB GVAR toolbox is available from https://sites.google.com/site/gvarmodelling/gvar-toolbox.

Apart from the easy-to-implement framework proposed by Robertson and Tallman (1999), a useful reference for conditional forecasting is Pesaran et al. (2007).

Finally, Huber and Ronchetti (2009) and Rousseeuw and Leroy (2003) are two major references within the robust statistic field of studies. Moreover, Atkinson and Riani (2000) and Atkinson et al. (2004) are the FS milestones. An interesting robust time series investigation is proposed in Lucas et al. (2011).

APPENDIX. ROBUST VECTOR ERROR CORRECTION MODEL: A FORWARD SEARCH APPROACH

Data may be corrupted by the presence of outliers. Models to fit data may also be inadequate to deal with atypical patterns. The identification of anomalous units and the immunization of data analysis against model failures are important aspects of modern statistics. A series of methods have been developed in the last few years. Among them, the FS deserves specific attention (Atkinson et al., 2004). Unlike other robust techniques, it aims to detect aberrant observations with suggestions for model enhancement by use of a graphical approach. The key idea of the FS is to monitor a given statistic as the model is fitted to subsets of increasing size (Atkinson and Riani, 2000). Therefore the key advantage of the FS is it is easy to interpret and implement. Indeed, it does not require any additional assumptions apart from those that characterize the model under analysis.

Three main phases characterize the process: initialization, progression, and monitoring of the search. Firstly, one needs to start from an outlier-free initial subset on which model parameters and residuals are estimated (i.e., initialization). As a second step, units are ranked according to a specified distance until all units are included in the subset (i.e., progression). The third task is to monitor some suitable quantities during the search (i.e., monitoring).

For the models analyzed in Section 2.3, parameters can be estimated via maximum likelihood in two steps (Johansen, 1988, 1991). Once the model parameters have been estimated, one may compute the fitting errors as follows:

$$\mathbf{e}_t = \Delta\mathbf{x}_t - \widehat{\Delta\mathbf{x}_t}, \tag{2.62}$$

where $\mathbf{e}_t$ is a $p \times 1$ vector.

These residuals constitute the key element of the FS. Indeed, let $S_*^{(m)}$ be the optimal subset of size m. Fitting errors, $\mathbf{e}_t(m^*)$, and standardized residuals, $r_t(m^*)$, are computed for all units $t = 1, \ldots, T$ by use of the vector of parameters $\hat{\mathbf{\Theta}}_*^{(m)}$ estimated on $S_*^{(m-1)}$. In terms of notation, $\mathbf{e}_t(m^*)$ stands for unit t vector of fitting errors at step m (obtained by our considering the parameter vector $\hat{\mathbf{\Theta}}_*^{(m)}$). Standardized residuals are obtained as follows:

$$r_t(m^*) = \sqrt{\mathbf{e}_t(m^*)'[\mathbf{s}^2(m^*)]^{-1}\mathbf{e}_t(m^*)}, \tag{2.63}$$

where $\mathbf{s}^2(m^*)$ is estimated

$$\mathbf{s}^2(m^*) = \frac{\sum_{t \in S_*^{(m)}} \mathbf{e}_t(m^*)\mathbf{e}_t(m^*)'}{m - 1}. \tag{2.64}$$

The subset $S_*^{(m)}$ is made up by the m units (not necessarily adjacent) with the lowest standardized residuals.

The parameter vector $\hat{\mathbf{\Theta}}_*^{(m+1)}$ is used to compute standardized residuals for the $m + 1$ step. Then the process continues until all units are included in the subset.

Let Eq. (2.65) denote the observation with the minimum standardized residual among those not in $S_*^{(m)}$:

$$x_{t_{min}} = \arg\min[r_t^*(m^*)] \ t \notin S_*^{(m)}. \tag{2.65}$$

The following quantity is monitored to test whether observation $x_{t_{min}}$ is an outlier:

$$r_{t_{min}}^*(m^*) = \sqrt{\mathbf{e}_{t_{min}}(m^*)'[\mathbf{s}^2(m^*)]^{-1}\mathbf{e}_{t_{min}}(m^*)}. \tag{2.66}$$

According to the above, the FS procedure can be summarized as follows:

- **Initialization.** The time series is split into q blocks. In line with the asymptotic analysis of Johansen and Nielsen (2012), the number of units of each block is chosen according to a fixed percentage of T (e.g., 10%). For each of the q blocks, the median of standardized residual is computed. The initial b-dimensional subset $(S_*^{(b)})$ is the one with the lowest median (Bellini, 2010). This is a generalization of the least median of squares criterion in regression (Rousseeuw, 1984).
- **Progression.** Once the initial subset has been chosen, standardized residuals are estimated according to Eq. (2.63). The optimal subset of size m (i.e., $S_*^{(m)}$) includes the m units with the lowest standardized residuals.

- **Monitoring.** Outlier detection is performed by the monitoring of the statistic introduced in Eq. (2.66). Atypical units are detected when the latter statistic crosses a given confidence threshold. On this subject, Atkinson and Riani (2006) studied the distribution of minimum deletion residuals in regression. They compared the empirical distribution with approximations based on truncated samples and order statistics. Riani et al. (2009) extended this analysis to the multivariate setting. An analytical approximation was found for standardized residuals. Bellini (2016) showed that this approach may be also applied to VEC models.

Lucas (1997) proposed a testing technique based on the pseudolikelihood estimation to reduce the effect of outliers. The parameters of Eq. (2.27) are estimated by use of the following Student T pseudolikelihood with v degrees of freedom:

$$\mathcal{L}(\Theta) = \prod_{t=1}^{T} \frac{\Gamma(\frac{v+p}{2})}{\Gamma(\frac{v}{2})|\pi v \Sigma|^{\frac{1}{2}}} \left(1 + \frac{\epsilon_t' \Sigma^{-1} \epsilon_t}{v} \right)^{-\frac{v+p}{2}}, \tag{2.67}$$

where Θ is the vector of unknown parameters, p denotes the number of variables, and Σ is the covariance matrix. The degrees of freedom v are chosen according to Franses and Lucas (1998) in the range $v = 3, \ldots, 7$. Lower values of v provide more protection against atypical observations but imply higher loss in power if there are no outliers.

As a useful by-product, weights are obtained for each observation. According to Franses and Lucas (1998), the Student T estimator for Eq. (2.27) can be considered as the Gaussian pseudo-maximum likelihood (PML) estimator for a weighted version of the data with weights given by

$$w_t = \left(\frac{v + p}{v + \epsilon_t' \Sigma^{-1} \epsilon_t} \right)^{\frac{1}{2}}. \tag{2.68}$$

Franses and Lucas (1998) suggest how to calculate their critical values to make inference on outliers. Under the assumption that ϵ_t are i.i.d. standard normally distributed, $\epsilon_t' \Sigma^{-1} \epsilon_t$ has a χ^2 distribution with p degrees of freedom. Let $\chi_p(0.005)$ denote the 0.5% value for the χ^2 distribution; weights are extraordinary small if $w_t \leq \left(\frac{v+p}{v+\chi_p(0.005)} \right)^{\frac{1}{2}}$. According to Franses and Lucas (1998), the PML analysis is conducted by use of $v = 5$ and by comparison of w_t with the critical value described above (for $v = 5$).

An empirical analysis is shown to better understand the entire process and compare FS and PML. Let us focus on a simulated time series. Firstly, no contamination is considered. The study starts without contamination, then additive, level shift, and innovation outliers are introduced. The entire analysis is done by our comparing the FS with the weighting scheme of Franses and Lucas (1998) implemented according to Bosco et al. (2010).

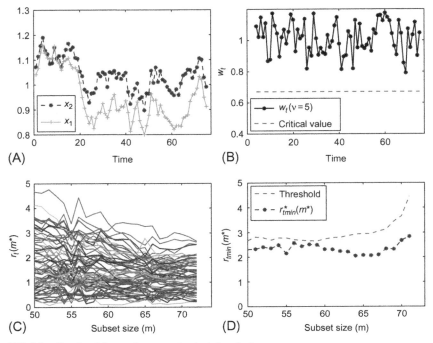

FIG. 2.9 Simulated time series: uncontaminated analysis.

The simulated analysis is based on the following bivariate (i.e., $p = 2$) model:

$$\begin{pmatrix} \Delta x_{1,t} \\ \Delta x_{2,t} \end{pmatrix} = \alpha\beta' \begin{pmatrix} x_{1,t-1} \\ x_{2,t-1} \end{pmatrix} + \Gamma_1 \begin{pmatrix} \Delta x_{1,t-1} \\ \Delta x_{2,t-1} \end{pmatrix} \qquad (2.69)$$

$$+ \Gamma_2 \begin{pmatrix} \Delta x_{1,t-2} \\ \Delta x_{2,t-2} \end{pmatrix} + \alpha\mu_0 + \begin{pmatrix} \epsilon_{1,t} \\ \epsilon_{2,t} \end{pmatrix},$$

where $t = 1,\ldots,75$. One path is simulated for each of the two time series $\mathbf{x}_t = (x_{1,t}, x_{2,t})'$ as highlighted in Fig. 2.9A. The three initial observations of the simulated time series are $(x_{1,t=1}, x_{2,t=1})' = (1.0420, 1.0720)'$, $(x_{1,t=2}, x_{2,t=2})' = (1.0670, 1.1150)'$, and $(x_{1,t=3}, x_{2,t=3})' = (1.1290, 1.1570)'$. The cointegration parameters are as follows:

$\alpha = (0.0113, 0.0177)'$, $\beta = (16.6046, -36.6281)'$,

$\Gamma_1 = \begin{pmatrix} -0.2313 & 0.4028 \\ -0.0029 & 0.1534 \end{pmatrix}$ and $\Gamma_2 = \begin{pmatrix} 0.1310 & 0.0790 \\ 0.0513 & 0.2103 \end{pmatrix}$, $\mu_0 = $

21.9292. The error terms, $\epsilon_{1,t}$ and $\epsilon_{2,t}$, are normally distributed with mean zero and covariance matrix $\Sigma = \begin{pmatrix} 0.0011 & 0.0010 \\ 0.0010 & 0.0013 \end{pmatrix}$.

Fig. 2.9A shows the simulated time series (75 observations). Fig. 2.9B highlights PML weights and shows that the critical value (i.e., $\nu = 5$) is not crossed. Thus no outlier is detected according to the PML method. For the FS, Fig. 2.9C and D outlines the evolution of residuals during the search. Fig. 2.9C shows the dynamic of standardized residuals $r_t(m^*)$ for each observation. Each line of this plot represents $r_t(m^*)$, for unit t $(t = 1, \ldots, T)$, as the subset size m increases. As a reminder, $r_t(m^*)$ stands for the standardized residuals for unit t at step m, obtained by use of the parameter vector $\hat{\Theta}_*^{(m)}$.[3] In other words, the subset size is represented on the horizontal axis, whereas the vertical axis describes the standardized residuals $r_t(m^*)$ for each unit. From left to right, at each subset size m, the evolution of residuals is shown through a line connecting $r_t(m^*)$ points. Fig. 2.9C shows that all lines are close to each other, meaning that there are no separate clusters of observations or atypical units. In Fig. 2.9D $r_{tmin}^*(m^*)$ is compared with the 99% confidence threshold; $r_{tmin}^*(m^*)$ is always below the 99% threshold, meaning that no atypical observation is detected. Both the PML and the FS do not reveal the presence of any outlier.

In what follows, three contamination schemes are considered. Additive, level shift, and innovation outliers are included in the above simulated time series.

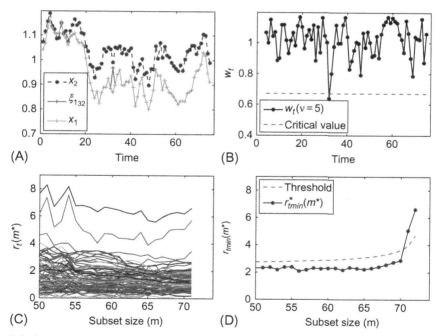

FIG. 2.10 Simulated time series: additive outliers.

---

3. It is worth mentioning that the symbol * relates to the parameter vector associated with the optimal subset (of size m) and not to the size of the subset.

Starting from additive outliers, the time series $\mathbf{x}_{1,t}$ is corrupted by the addition of 0.1 to the observation at time 32 (which represents more than 10% of its original value). Fig. 2.10A pinpoints unit $x_{1,32}$ (indicated with $\xi_{1,32}$). Two anomalous movements occur in the first time series. Fig 2.10B shows that both unit 32 and unit 33 are underweighted. However, only observation 32 crosses the critical value and can be considered an outlying unit. For the FS, the effect of the contamination is evident in Fig. 2.10C, where the evolution of $r_t(m^*)$ is shown. Two observations, units 32 and 33, move separately from all other units. In Fig. 2.10D the comparison of $r^*_{tmin}(m^*)$ against the 99% confidence threshold shows that units 32 and 33 are outliers.

The analysis continues by introducing a level shift on $\mathbf{x}_{2,t}$ time series. In particular, as emphasized in Fig. 2.11A, units $x_{2,\{65,\ldots,75\}}$ are subject to a shift of 0.15. Fig. 2.11B shows that unit 65 crosses the critical value, pointing out the level shift. The weights assigned to units from 66 to 75 do not cross the critical value. For the FS, Fig. 2.11C highlights two separate clusters of trajectories from the very beginning until the last few steps of the search. All contaminated units have higher $r_t(m^*)$ during the search, but in the last few steps there is a masking phenomenon. As emphasized in Fig. 2.11D, the only observation that is identified as an outlier is unit 65. Therefore the joint analysis of Fig. 2.11C and D allows us to correctly identify all contaminated units.

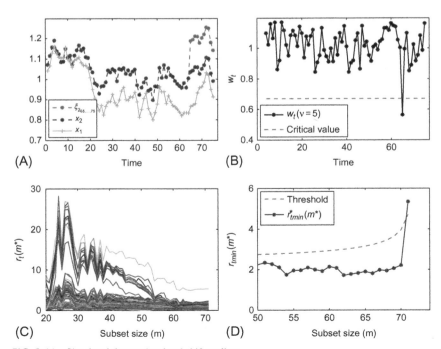

FIG. 2.11 Simulated time series: level shift outliers.

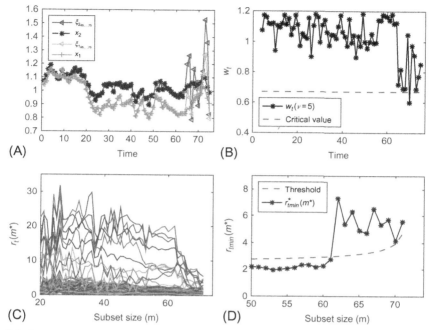

FIG. 2.12 Simulated time series: innovation outliers.

The last contamination scheme is associated with innovation outliers. Units $x_{\{1,2\},\{65,...,75\}}$ are contaminated by the application of a random shock from a normal variable with mean 0 and variance 0.12. Fig. 2.12A highlights innovation outliers $x_{\{1,2\},\{65,...,75\}}$ indicated with $\xi_{\{1,2\},\{65,...,75\}}$. Fig. 2.12B shows that only one of the last units crosses the critical value. The weights assigned to other contaminated observations do not cross the critical value and show that no further observations are detected as outliers. Fig. 2.12C recognizes that innovation outliers are slightly masked in the last steps of the search. Fig. 2.12D, comparing $r^*_{tmin}(m^*)$ with the 99% confidence threshold, emphasizes that all contaminated units (except for the one that is very close to the original pattern) are correctly detected.

Fig. 2.13 additionally helps the analysis by showing units not belonging to the subset at each step of the search in the innovation outliers analysis. In particular, Fig. 2.13A indicates with a filled circle units not belonging to the subset. In Fig. 2.13B the focus is on the end of the search. All contaminated units enter the subset in the last steps of the search. Fig. 2.13C and D focuses on contaminated units. Fig. 2.13C shows the evolution of $r_t(m^*)$ for the corrupted observations. Fig. 2.13D highlights the major contribution of $x_{2,t}$ series to the overall outlying pattern.

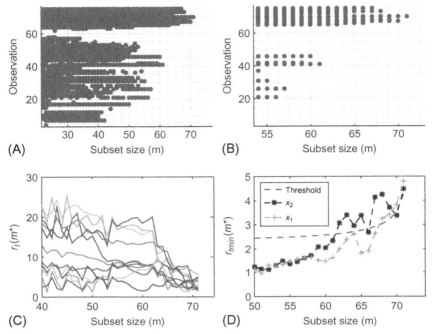

FIG. 2.13 Simulated time series with innovation outliers: analysis of units not belonging to the subset.

Bellini (2016) performed additional Monte Carlo experiments and showed that the FS outperforms the PML in the case of a cluster of atypical observations and when innovation outliers characterize the data under analysis.

Despite the increasing importance of robust statistics, scepticism still characterizes the use of these methods in time series analysis. The main sources of scepticism can be summarized as follows:

(i) There are a plethora of semiparametric and nonparametric techniques for outlier analysis that can be used to deal with contaminated data.

(ii) Outlier robust procedures are not standardized enough, leaving the user too much freedom in choosing parameters to be used in the analysis (see, e.g., M-estimators).

(iii) The computational effort required to compute robust estimators is higher than for traditional methods.

(iv) Time series analysts tend to consider outlier robust statistics as a black box where observations are manipulated or weighted to fit the data.

With regard to the first issue, one needs to bear in mind that semiparametric and nonparametric techniques generally require many more observations to work reasonably well. In addition, they are not free from outlier and data structure problems. In other words, even when applying these methods, the researcher has to be cautious about atypical observations.

The second argument points out that some robust approaches allow the user to choose among different functional forms or set of parameters to run the estimation. The FS is not sensitive to this issue because it relies on the standard maximum likelihood estimation without requiring any further assumption on the model to be fitted.

When one is dealing with real data, the need to speed up computations is particularly significant. The FS and some other robust methods are subject to intensive computational efforts. However, fast procedures to obtain confidence thresholds are obtained without the need for Monte Carlo simulations (Riani et al., 2009).

Finally, the FS is not a black-box procedure, and manipulation or weighting schemes are not applied. On the contrary, the FS monitors suitable statistics as the subset size increases. A battery of plots are available for the researcher to infer outliers by allowing the researcher to decide what to do with these units (Nielsen, 2004).

EXERCISES

Exercise 2.1 Let us consider the following AR(2) process:

$$x_t = 0.5 + 0.8x_{t-1} - 0.1x_{t-2} + \epsilon_t.$$

- Simulate 50 observations of this process by use of the MATLAB Econometrics Toolbox and plot the corresponding graph.
- Estimate the parameters of the model by use of the MATLAB function *estimate*.

Exercise 2.2 Exercise 2.1 is the starting point for the following analysis.

- Fit an ordinary least squares regression model on the time series of Exercise 2.1 by use of MATLAB.

Exercise 2.3 Let us consider the following ARMA(1,1) process:

$$x_t = 0.8x_{t-1} + 0.2\epsilon_{t-1} + \epsilon_t.$$

- Simulate 50 observations of the above process by use of R.
- Plot the autocorrelation and partial-autocorrelation functions.
- Fit the simulated process.

Exercise 2.4 The file Chap2Ex4.xlsx contains Italian time series spanning from 1993 to 2010. The focus is on the following quarterly variables: default rate (*DR*), gross domestic product (*GDP*), unemployment rate (*UR*), real estate price index (*RE*), exchange rate of the euro against the dollar (*ER*), the 3-month Euribor (*EUR3M*), and the 10-year interest rate swap (*IRS*10). The reader is invited to conduct the following analysis with R:

- Perform the Johansen trace test.
- Check residual properties.
- Run the impulse response analysis after having converted the VEC model into a VAR model.

Exercise 2.5 Let us consider the time series and MATLAB GVAR toolbox available from https://sites.google.com/site/gvarmodelling/gvartoolbox. The reader is invited to replicate the analysis described in Section 2.4.2.

Solutions are available at www.tizianobellini.com.

REFERENCES

Atkinson, A.C., Riani, M., 2000. Robust Diagnostic Regression Analysis. Springer, New York.

Atkinson, A.C., Riani, M., 2006. Distribution theory and simulations for tests of outliers in regression. J. Comput. Graph. Stat. 15, 460–476.

Atkinson, A.C., Riani, M., Cerioli, A., 2004. Exploring Multivariate Data with the Forward Search. Springer, New York.

Bellini, T., 2010. Detecting atypical observations in financial data: the forward search for elliptical copulas. Adv. Data Anal. Classif. 4, 287–299.

Bellini, T., 2016. The forward search interactive outlier detection in cointegrated VAR analysis. Adv. Data Anal. Classif. 10, 351–373.

Bosco, B., Parisio, L., Pelagatti, M., Baldi, F., 2010. Long-run relations in European electricity prices. J. Appl. Econ. 25, 805–832.

Box, G.E.P., Jenkins, G.M., 1976. Time Series Analysis: Forecasting and Control. Holden-Day, San Francisco.

Castren, O., Dees, S., Zaher, F., 2010. Stress-testing Euro area corporate default probabilities using a global macroeconomic model. J. Financ. Stab. 6, 64–74.

Dees, S., Di Mauro, F., Pesaran, M., Smith, L.V., 2007. Exploring the international linkages of the Euro area: a global VAR analysis. J. Appl. Econ. 22, 1–38.

Di Mauro, F., Pesaran, M., 2013. The GVAR Handbook. Oxford University Press, Oxford.

Dickey, A., Fuller, W.A., 1981. Likelihood ratio statistics for autoregressive time series with unit root. Econometrica 49, 1057–1072.

Doan, T., Litterman, R.B., Sims, A., 1984. Forecasting and conditional projections using realistic prior distributions. Econ. Rev. 3, 1–100.

Franses, P., Lucas, A., 1998. Outlier detection in cointegration analysis. J. Bus. Econ. Stat. 16, 459–468.

Hamilton, J., 1994. Time Series Analysis. Princeton University Press, Princeton.

Huber, P., Ronchetti, E., 2009. Robust Statistics. Wiley, Hoboken.

Johansen, S., 1988. Statistical analysis of cointegration vectors. J. Econ. Dyn. Control 12, 231–254.

Johansen, S., 1991. Estimation and hypothesis testing of cointegration vectors in Gaussian vector autoregressive models. Econometrica 59, 1551–1580.

Johansen, S., 1996. Likelihood-based Inference in Cointegrated Vector Autoregressive Models. Oxford University Press, Oxford.

Johansen, S., Nielsen, B., 2012. Asymptotic analysis of the forward search. Working Paper.

Juselius, K., 2006. The Cointegrated VAR model: Methodology and Applications. Oxford University Press, Oxford.

Kydland, F.E., Prescott, E.C., 1982. Time to build and aggregate fluctuations. Econometrica 50, 1345–1370.

LeSage, J.P., 1999. Applied Econometrics using MATLAB. University of Toledo. http://www.spatial-econometrics.com/html/mbook.pdf.

Lucas, A., 1997. Cointegration testing using pseudolikelihood ratio tests. Econ. Theory 13, 149–169.

Lucas, A., Fransen, P., Van Dijk, D., 2011. Outlier Robust Analysis of Economic Time Series. Oxford University Press, Oxford.

Lutkepohl, H., 2005. New Introduction to Multiple Time Series Analysis. Springer, Berlin.

Nielsen, H.B., 2004. Cointegration analysis in the presence of outliers. Econ. J. 7, 249–271.

Pesaran, M., Shin, Y., 1998. Generalized impulse response analysis in linear multivariate models. Econ. Lett. 58, 17–29.

Pesaran, M., Schuermann, T., Weiner, S., 2004. Modeling regional interdependencies using a global error-correcting macroeconometric model. J. Bus. Econ. Stat. 22, 129–162.

Pesaran, M., Smith, L.V., Smith, R.P., 2007. What if the UK or Sweden had joined the Euro in 1999? An empirical evaluation using a global VAR. Int. J. Finance Econ. 12, 55–87.

Pfaff, B., 2008. Analysis of Integrated and Cointegrated Time Series with R. Springer, New York.

Riani, M., Atkinson, A.C., Cerioli, A., 2009. Finding an unknown number of multivariate outliers. J. R. Stat. Soc. Ser. B 71, 201–221.

Robertson, J.C., Tallman, E.W., 1999. Vector autoregressions: forecasting and reality. Econ. Rev. 84 (1), Federal Reserve Bank of Atlanta.

Rousseeuw, P., 1984. Least median of squares regression. J. Am. Stat. Soc. 79, 871–880.

Rousseeuw, P., Leroy, A., 2003. Robust Regression and Outlier Detection. Wiley, Hoboken.

Chapter 3

Asset and Liability Management, and Value at Risk

Chapter Outline

Asset and liability management (ALM) was developed in the mid-1970s with the aim of fostering bank performance in the face of high and volatile interest rates. Its objective being to coordinate the functions affecting interest-bearing assets and interest-burdened liabilities, ALM is also a model archetype for the analysis of margin, value, and liquidity risks.

This chapter introduces Bank Alpha, a stylized international commercial bank used as an illustrative example throughout the book. A brief description of its balance sheet and profit and loss paves the way to the ALM analysis.

Firstly, the study focuses on the net interest income (NII) computed as the difference between interest revenue and expenses over a given time horizon (e.g., 1 year). Given the link between NII and interest rates, the investigation

Stress Testing and Risk Integration in Banks. http://dx.doi.org/10.1016/B978-0-12-803590-0.00003-5

of an affine term structure model shows how to assess margin fluctuations due to macroeconomic changes.

As part of the stress testing and risk integration process, the next step of the journey proposes a quick overview of value at risk (VaR) methods. Three main approaches are considered: variance-covariance, Monte Carlo, and historical simulations. A methodological description is presented, followed by its regulatory implementation.

Finally, the ALM mechanism is used to analyze the mismatch between cash outflows and inflows. Bank Alpha serves to illustrate some of the main liquidity issues experienced during the 2007–09 crisis.

KEY ABBREVIATIONS AND SYMBOLS

$A_{i,t}$	customer i asset at time t
$A_{i,ra,t}$	customer i asset residual amount, ra, at time t
$A_{i,cf,t}$	customer i asset cash flow, cf, at time t
ΔA	total asset value change
$C_{i,t}$	customer i facility coupon at time t
$C_{j,t}$	liability j coupon at time t
$\Delta C_{(\cdot,t)}$	coupon rate variation at time t
$DD_{i,t}$	customer i facility time period, DD, from t to the next repricing date or end of the period
$DD_{j,t}$	liability j time period, DD, from t to the next repricing date or end of the period
DD_l	year fraction from l to the end of the holding period, h
$\overline{DD}_{A,l}$	bucket l asset A average period from l and end of h
$Gap_{rep,l}$	bucket l repricing, rep, gap
$Gap_{liq,ll}$	bucket ll liquidity, liq, gap
$IA_{i,h}$	customer i asset interest over the holding period h
$IL_{j,h}$	liability j interest over the holding period h
$L_{j,t}$	liability j at time t
$L_{j,ra,t}$	liability j residual amount, ra, at time t
$L_{j,cf,t}$	liability j cash flow, cf, at time t
ΔL	total liability value change
NII_h	net interest income over the holding period h
$NII_{h,BG}$	net interest income over the holding period h considering balance growth, BG, assumptions
ΔNII_h	net interest income variation, Δ, over h
$\Delta NII_{h,\Delta}$	net interest income variation over h due to shocked macro variables $\mathbf{x}_\Delta$
PV	present value
PVA_i	asset i present value
PVL_j	liability j present value

$R_{t,0}$ risk-free interest rate at t ongoing at $t = 0$
ΔR_t time t risk-free interest rate variation
$\mathbf{x}_t$ vector of macroeconomic variables at time t
$\mathbf{x}_{t,\Delta}$ shocked vector of macroeconomic variables

3.1 INTRODUCTION

A balance sheet analysis provides an intuitive representation of an asset and liability structure. In this regard, let us consider a bank with N assets $\mathbf{A}_t = (A_{1,t},\ldots,A_{N,t})'$ and M liabilities $\mathbf{L}_t = (L_{1,t},\ldots,L_{M,t})'$ at a given time t. Each exposure has a given size, maturity, coupon rate, time to repricing, and so on.

In line with the above, Table 3.1 introduces Bank Alpha as an illustrative example of a stylized international commercial bank. At t_0, the stress testing and risk integration starting point, the assets comprise $70 billion of loans, $14 billion of securities, $8 billion cash, and $8 billion of other instruments. Assets are reported net of provisions. From a liability perspective, deposits account for $70 billion, while other liabilities, including bonds and acceptances, sum up to $17 billion. Subordinated debts account for $4 billion and noncontrolling interest for $2 billion. Finally, shareholder equity is reported as $7 billion. For simplicity, off-balance sheet operations are not considered.

If the balance sheet represents a snapshot of a firm at a given point in time, the profit and loss statement displays the movements occurring over a given time horizon (e.g., 1 year). Table 3.2 highlights that NII, noninterest revenue, and noninterest expenses are added up to obtain the preprovisioning net revenue. Loan impairment charges and taxes are the additional ingredients to compute the net profit (loss). Table 3.2 summarizes Bank Alpha's profit and loss over the 1-year period $[t_0, t_1]$, given the state of the economy at t_0. In this regard, one needs to bear in mind that the $1.06 billion net profit is threatened by potential economic downturns. The role of a stress testing process is to highlight these possible changes due to adverse scenarios.

The first goal pursued in this chapter is to highlight the margin at risk. According to Bank Alpha's profit and loss representation, Section 3.2 concentrates on movements affecting the $2.50 billion NII shown in Table 3.2. An ALM framework is required to explore these fluctuations. Indeed, changes may affect both interest revenue and interest expenses. Therefore assets as well as liabilities require precise scrutiny.

Macroeconomic shocks do not affect only a bank's margin. They cause variations in the value of assets and liabilities. Despite the potential interest for a full fair value estimation of both sides of the balance sheet, accounting principles commonly restrict such assessment to the trading book. For this reason, in Section 3.3 specific attention is devoted to the market VaR. In terms of Bank

TABLE 3.1 Bank Alpha's Balance Sheet at t_0 ($ Billions)

Balance Sheet

Assets				Liabilities		
Cash resources			8.00	Deposits		70.00
	Cash	2.00		Non-interest bearing	5.00	
	Interest-bearing deposits	6.00		Interest-bearing	65.00	
Securities			14.00	Other liabilities		17.00
	Trading account	3.00		Acceptances	3.00	
	Available for sale	7.00		Bonds	11.00	
	Held to maturity	4.00		Other residual liabilities	3.00	
Loans			70.00	Subordinated debts		4.00
	Corporate loans	25.00		Noncontrolling interests		2.00
	Retail loans	25.00		Shareholder equity		7.00
	Real estate loans	20.00		Common shares	4.00	
Other assets			8.00	Preferred shares	1.00	
				Retained earnings	2.00	
Total assets			100.00	Total liabilities		100.00

TABLE 3.2 Bank Alpha's Profit and Loss Over the Period $[t_0, t_1]$, Given the State of the Economy at t_0 ($ Billions)

Profit and Loss	
Net interest income	2.50
Noninterest revenue	2.51
Noninterest expenses	−3.01
Preprovisioning net revenue	2.00
Loan impairment charges	−0.55
Tax	−0.39
Net profit/loss	1.06

Alpha's balance sheet, the study targets a potential reduction of the $3.00 billion trading account portfolio value.

The study of liquidity misalignments characterizes the final part of this chapter. As per the margin at risk analysis, an ALM mechanism applies to this study. Section 3.4 relies on a mapping process of cash inflows and outflows to verify whether a bank is subject to potential liquidity issues.

From a toolkit perspective, the ALM framework plays a central role throughout the entire chapter. However, some additional tools enrich the analysis. Firstly, an optimization procedure is required, within the margin at risk analysis, to assess the interest rate sensitivity to economic changes. Then a Kalman filter system is used to fit the term structure of interest rates based on the models of Vasicek (1977) and Cox et al. (1985). Finally, additional statistical techniques allow us to conduct the VaR analysis by distinguishing among variance-covariance, Monte Carlo, and historical simulation approaches.

3.2 MARGIN AT RISK

In this section we investigate the margin at risk as the difference between interest revenue and interest expenses (i.e., NII). Let us start from a generic (interest rate sensitive) asset $A_{i,t,ra}$, where the notation identifies the exposure referred to customer i at time t. The subscript ra symbolizes the residual amount. The following equation shows how to compute the interest revenue on the above asset (IA) over a given holding period h (e.g., 1 year):

$$IA_{i,h} = \sum_{t=0}^{h} \mathbb{1}_{i,rep,t} A_{i,ra,t} C_{i,t} DD_{i,t}, \tag{3.1}$$

where $C_{i,t}$ is the coupon rate at time t and $DD_{i,t}$ represents the time period to the next repricing date (within the h period) or to the end of the holding period (if h occurs before the next repricing date); $\mathbb{1}_{i,rep,t}$ assumes the value 1 at $t = 0$ or when a repricing (resetting of interest rate) occurs[1]:

$$\mathbb{1}_{i,rep,t} = \begin{cases} 1 & \text{for} \quad t = 0 \quad \text{or} \quad t = \text{repricing}, \\ 0 & \text{otherwise}. \end{cases}$$

Example 3.1 shows how Eq. (3.1) works in practice.

Example 3.1 Interest Revenue on a Given Asset Over a 1-Year Holding Period ($IA_{i,h}$)

Let us consider a customer with a $10 million bullet loan exposure having a 3-year maturity and floating rate indexed to the LIBOR. The interest rate formula is LIBOR + spread, where the contractual annual spread is 1.0%. This loan pays interest on a 6-month basis.

The LIBOR used for the first period is 2.00% (on an annual basis). Given the term structure of interest rates at t_0, the market expectation for the LIBOR at time $t = 6$ months is 2.50% (on an annual basis). Table 3.3 summarizes the interest revenues computed over a 1-year holding period.

TABLE 3.3 Asset Interest Revenues for a $10 Million Bullet Loan ($ Millions)

t^a	$\mathbb{1}_{i,rep,t}$	$A_{i,ra,t}$	$C_{i,t}$ (%)	$DD_{i,t}{}^b$	$\mathbb{1}_{i,rep,t} A_{i,ra,t} C_{i,t} DD_{i,t}$
0.00	1	10.00	3.00	180/360	0.150
...	0	10.00	3.00	...	0.000
...	0	10.00	3.00	...	0.000
0.50	1	10.00	3.50	180/360	0.175
...	0	10.00	3.50	...	0.000
...	0	10.00	3.50	...	0.000
Total					0.325

^{a} Over a continuous interval.
^{b} In days over 360.

All in all, the indicator function $\mathbb{1}_{i,rep,t}$ assumes the value 1 only at $t = 0$ and $t = 6$ months. According to the term structure of interest rates at t_0, the interest revenue over the 1-year holding period (i.e., $IA_{i,h}$) is $0.325 million.

---

1. For simplicity, this formalization represents the usual case where floating interest rates change simultaneously with amortization repayments.

On the liability side, $IL_{j,h}$ represents the interest due to liability j for the holding period h. Interest expenses result from what follows:

$$IL_{j,h} = \sum_{t=0}^{h} \mathbb{1}_{j,rep,t} L_{j,ra,t} C_{j,t} DD_{j,t},$$ (3.2)

(3.3)

where, other things have meanings similar to those in Eq. (3.1), $L_{j,ra,t}$ stands for the jth liability residual amount (ra) at time t.

The NII for a portfolio of N assets and M liabilities is computed as follows:

$$NII_h = \sum_{t=0}^{h} \left(\sum_{i=1}^{N} \mathbb{1}_{i,rep,t} A_{i,ra,t} C_{i,t} DD_{i,t} - \sum_{j=1}^{M} \mathbb{1}_{j,rep,t} L_{j,ra,t} C_{j,t} DD_{j,t} \right).$$ (3.4)

Example 3.2 shows how Eq. (3.4) works in practice by considering a one-asset, one-liability portfolio.

Example 3.2 NII for a One-Asset, One-Liability Portfolio

A portfolio has one asset $(A_{i,t})$ and one liability $(L_{j,t})$ sensitive to interest rates. Additionally, equity funds the imbalance between $A_{i,t}$ and $L_{j,t}$. Let us estimate the NII over a 1-year holding period h by use of the following additional information:

- $A_{i,t}$ has the following characteristics:
  - At inception, Jan. 1, the residual amount (A_{i,ra,t_0}) is \$1 million.
  - The residual maturity is 5 years.
  - A floating interest is due on Jun. 30 and Dec. 31. The interest rate formula is LIBOR + spread, where the contractual annual spread is 2.0%. On Jan. 1, the reference LIBOR is 3% (on an annual basis). According to the corresponding term structure, the LIBOR will be 2% (on an annual basis) in 6 months. A \$0.1 million principal repayment is due on Jun. 30 and Dec. 31.
- $L_{j,t}$ has the following characteristics:
  - At inception the bullet bond residual amount $(L_{j,ra,t})$ is \$0.8 million.
  - The residual maturity is 2 years.
  - The fixed coupon rate, $C_{j,t}$, is 3% (on an annual basis). Interest is paid annually on Dec. 30, while capital is reimbursed at maturity.
- The initial \$0.2 million imbalance between $A_{i,t}$ and $L_{j,t}$ is funded with equity for which neither interest expenses nor maturity applies.

Fig. 3.1 summarizes the steps to compute the 1-year asset interest, IA. For the first 6 months the interest is \$0.025 million. For the next 6 months the interest is \$0.018 million. For the entire period the interest is \$0.043 million.

Table 3.4 summarizes the key components of Eq. (3.1) applied in the case under analysis.

(Continued)

Example 3.2 NII for a One-Asset, One-Liability Portfolio—cont'd

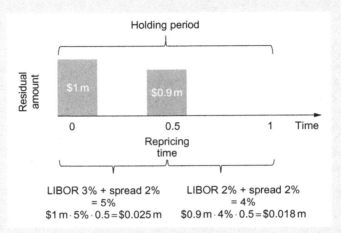

FIG. 3.1 Interest computation scheme for $1 million of amortization assets over a 1-year holding period.

TABLE 3.4 Asset Interest Revenues for a $1 Million Amortizing Loan ($ Millions)

t^a	$\mathbb{1}_{i,rep,t}$	$A_{i,ra,t}$	$C_{i,t}$	$DD_{i,t}^{\ b}$	$\mathbb{1}_{i,rep,t}A_{i,ra,t}C_{i,t}DD_{i,t}$
0.00	1	1.00	5.00%	180/360	0.025
...	0	1.00	5.00%	...	−
...	0	1.00	5.00%	...	−
0.50	1	0.90	4.00%	180/360	0.018
...	0	0.90	4.00%	...	−
...	0	0.90	4.00%	...	−
Total					0.043

[a] Over a continuous interval.
[b] In days over 360.

On the liability side, the 1-year interest of $0.024 million is computed as the product of the residual amount of $0.8 million and the fixed interest rate of 3%. Finally, NII of $0.019 million is derived as the difference between the asset interest of $0.043 million and the liability interest of $0.024 million.

When projecting NII, one needs to bear in mind that some financial instruments may expire within the holding period (e.g., maturity less than h) or reduce because of amortization. In contrast, others may not have a contractual maturity (e.g., current account). On the subject of balance sheet projection, two main approaches may be considered for stress testing. On the one hand, the EBA (2016), for example, uses a static assumption. On the other hand, the BOE (2015) and the FRB (2015), among others, base their exercises on a dynamic balance.

Eq. (3.4) makes the implicit assumption that $A_{i,ra,t}$ and $L_{j,ra,t}$ reduce because of repayments. Nonetheless, new operations continuously occur in banking business. A rolling scheme usually applies when one is dealing with short-term projections (i.e., a 1-year holding period). In other words, at expiry, a financial instrument is replaced with another with the same characteristics. The goal is now to extend Eq. (3.4) to include business dynamics. Growth assumptions need to be defined for both assets and liabilities as detailed in the following more comprehensive equation:

$$
\begin{aligned}
NII_{h,BG} = \ & \sum_{t=0}^{h} \left(\sum_{i=1}^{N} \mathbb{1}_{i,rep,t} A_{i,ra,t} C_{i,t} DD_{i,t} \right. \\
& + \sum_{iBG=1}^{N_{BG}} \mathbb{1}_{iBG,rep,t} A_{iBG,ra,t} C_{iBG,t} DD_{iBG,t} \Big) \\
& - \sum_{t=0}^{h} \left(\sum_{j=1}^{M} \mathbb{1}_{j,rep,t} L_{j,ra,t} C_{j,t} DD_{j,t} \right. \\
& + \sum_{jBG=1}^{M_{BG}} \mathbb{1}_{jBG,rep,t} L_{jBG,ra,t} C_{jBG,t} DD_{jBG,t} \Big),
\end{aligned}
\tag{3.5}
$$

where $NII_{h,BG}$ includes the balance sheet growth (indexed by BG). For both assets and liabilities, the balance sheet growth is usually positive. However $A_{iBG,ra,t}$ and $L_{jBG,ra,t}$ may include components with a negative sign. Moreover, the above-mentioned rolling scheme is a special case of balance sheet growth where each operation is replaced by another with the same characteristics.

Table 3.5 summarizes the rolling scheme by highlighting Bank Alpha's 1-year interest computation over the period $[t_0, t_1]$. It is worth noting the distinction between financial instruments sensitive and insensitive to interest rates. As an example, cash physically available does not generate interest. Similarly, shares do not produce interest but produce other forms of earnings. The $2.50 billion NII corresponds to what is shown in Table 3.2.

Balance sheet projections are presented in Chapter 5, where Bank Alpha's growth assumption is widely discussed together with its impact on the NII stress test.

In the next section a mapping process is introduced with the aim of facilitating margin at risk and sensitivity analysis. For simplicity, the rolling assumption is used throughout this chapter.

3.2.1 Margin at Risk Estimation

The first step to estimate the margin at risk is to distinguish between interest-sensitive and interest-insensitive financial instruments. Secondly, sensitive

TABLE 3.5 Bank Alpha's Net Interest Income Computed Under the Rolling Scheme Over the 1-year Period $[t_0, t_1]$ ($ Billions)

		Assets			
		Insensitive	A_{ra}	$\overline{C}_{(\cdot)}$	IA_h
Cash resources			6.00	3.00%	0.18
	Cash	2.00			
	Interest-bearing deposits		6.00	3.00%	0.18
Securities			13.00	3.38%	0.44
	Trading account	1.00	2.00	3.50%	0.07
	Available for sale		7.00	3.29%	0.23
	Held to maturity		4.00	3.50%	0.14
Loans			70.00	5.21%	3.65
	Corporate loans		25.00	5.88%	1.47
	Retail loans		25.00	5.16%	1.29
	Real estate loans		20.00	4.44%	0.89
Other assets		8.00			
Total sensitive assets			89.00	4.80%	4.27

		Liabilities			
		Insensitive	L_{ra}	$\overline{C}_{(\cdot)}$	IL_h
Deposits			65.00	2.17%	1.41
	Non-interest bearing	5.00			
	Interest bearing		65.00	2.17%	1.41
Other liabilities			17.00	1.65%	0.28
	Acceptances		3.00	0.67%	0.02
	Bonds		11.00	2.27%	0.25
	Other residual liabilities		3.00	0.33%	0.01
Subordinated debts			4.00	2.00%	0.08

TABLE 3.5 Bank Alpha's Net Interest Income Computed Under the Rolling Scheme Over the 1-year Period $[t_0, t_1]$ ($ Billions) — cont'd

		Liabilities			
		Insensitive	L_{ra}	$\overline{C}_{(\cdot)}$	IL_h
Noncontrolling interests		2.00			
Shareholder equity					
	Common shares	4.00			
	Preferred shares	1.00			
	Retained earnings	2.00			
Total sensitive liabilities			86.00	2.06%	1.77
Net interest income					2.50

Notes: From a notation point of view, $\overline{C}_{(\cdot)}$ in this table represents the average coupon applied over the 1-year holding period.

instruments are mapped on a time grid according to their repricing scheme. Then gaps are computed as the difference between assets and liabilities for each repricing bucket. Finally, the margin at risk derives from an interest rate variation applied to each gap (weighted by the time remaining to the end of the holding period h). In what follows, these steps are explored in more detail by our focusing on Bank Alpha's ALM:

- **Split between interest-sensitive and interest-insensitive financial instruments.** The first step of the process is to separate interest-sensitive from interest-insensitive financial instruments. Assets and liabilities are then aggregated by facility type, pivotal interest rates, and so on. As an example, Table 3.6 shows Bank Alpha's structure. In this case, sensitive assets accounts for $89 billion (i.e., $43.5 billion floating interest rate and $45.5 billion fixed interest rate), while assets worth $11 billion are insensitive. On the liability side, deposits and other interest rate-sensitive instruments sum up to $86 billion (i.e., $50 billion floating interest rate and $36 billion fixed interest rate). Insensitive liabilities add up to $14 billion.

TABLE 3.6 Bank Alpha's Assets and Liabilities Interest Rate Sensitivity Classification ($ Billions)

Assets		Floating	Fixed	Insensitive	Total
Cash resources					8.00
	Cash			2.00	
	Interest-bearing deposits		6.00		
Securities					14.00
	Trading account	1.00	1.00	1.00	
	Available for sale	4.00	3.00		
	Held to maturity	1.00	3.00		
Loans					70.00
	Corporate loans	15.00	10.00		
	Retail loans	12.50	12.50		
	Real estate loans	10.00	10.00		
Other assets				8.00	8.00
Total assets		43.50	45.50	11.00	100.00
Liabilities		**Floating**	**Fixed**	**Insensitive**	**Total**
Deposits					70.00
	Non-interest bearing			5.00	
	Interest bearing	40.00	25.00		
Other liabilities					17.00
	Acceptances	3.00			
	Bonds		11.00		
	Other residual liabilities	3.00			
Subordinated debts		4.00			4.00

TABLE 3.6 Bank Alpha's Assets and Liabilities Interest Rate Sensitivity Classification ($ Billions)—cont'd

		Liabilities			
		Floating	Fixed	Insensitive	Total
Noncontrolling interests				2.00	2.00
Shareholder equity					7.00
	Common shares			4.00	
	Preferred shares			1.00	
	Retained earnings			2.00	
Total liabilities		50.00	36.00	14.00	100.00

- **Time grid bucketing.** Repricing buckets are identified as time aggregation intervals. Each financial instrument notional residual amount (i.e., $A_{i,ra,t}$ and $L_{j,ra,t}$) is mapped on a repricing bucket (i.e., the bucket where a repricing occurs or a financial instrument expires). On this subject, Table 3.7 shows Bank Alpha's assets and liabilities across the following monthly buckets: [0–1M], (1M–3M], (3M–6M], (6M–12M], (>12M).[2] An important distinction arises between floating and fixed interest rate instruments. As an example, a financial instrument repricing in 2 months is allocated to the (1M–3M] bucket. Another repricing in 7 months ends up in the (6M–12M] bucket, and so on. In contrast, the outstanding balance referred to a fixed interest rate instrument is allocated to the bucket of its maturity (or repayment). In both cases the rolling assumption is used (i.e., when a repayment occurs, a new operation with the same financial characteristics originates).

- **Gap analysis**. Once notional residual amounts have been mapped to buckets, the gap between assets and liabilities is computed as follows:

$$Gap_{rep,l} = \sum_{i=1}^{N_l} \mathbb{1}_{i,rep,t_{rep} \in l} A_{i,ra,t_{rep}} - \sum_{j=1}^{M_l} \mathbb{1}_{j,rep,t_{rep} \in l} L_{j,ra,t_{rep}}, \qquad (3.6)$$

where $\mathbb{1}_{i,rep,t \in l}$ is 1 when a repricing occurs within the bucket as detailed below:

---

2. *M* stands for *month*, square brackets include extremes, and parentheses do not include extremes.

TABLE 3.7 Bank Alpha's Financial Instrument Mapping ($ Billions)

		Assets						
		[0–1M]	(1M–3M]	(3M–6M]	(6M–12M]	(>12M)	Insensitive	Total
Cash resources								8.00
	Cash						2.00	
	Interest-bearing deposits				6.00			
Securities							1.00	14.00
	Trading account	1.00	1.00					
	Available for sale		4.00		3.00			
	Held to maturity			1.00	0.60	2.40		
Loans								70.00
	Corporate loans	2.00	3.00	10.00	4.00	6.00		
	Retail loans			12.50	6.25	6.25		
	Real estate loans				11.14	8.86		
Other assets							8.00	8.00
Total assets		3.00	8.00	23.50	30.99	23.51	11.00	100.00

Liabilities

		[0–1M]	(1M–3M]	(3M–6M]	(6M–12M]	(> 12M)	Insensitive	Total
Deposits	Non–interest bearing						5.00	70.00
	Interest bearing	62.50		2.50				
Other liabilities	Acceptances				3.00			17.00
	Bonds				6.00	5.00		
	Other residual liabilities				3.00			
Subordinated debts					4.00			4.00
Noncontrolling interests							2.00	2.00
Shareholder equity	Common shares						4.00	7.00
	Preferred shares						1.00	
	Retained earnings						2.00	
Total liabilities		62.50	8.00	2.50	16.00	5.00	14.00	100.00
Gap		−59.50	8.00	21.00	14.99	18.51		

M, Month(s).

$$\mathbb{1}_{i,rep,t\in l} = \begin{cases} 1 & \text{for} \quad t_{rep} \in l, \\ 0 & \text{otherwise.} \end{cases}$$

The last row of Table 3.7 highlights Bank Alpha's gaps. These gaps feed the next step of the process to summarize the impact of interest rate changes on the bank's margin. It is worth noting that Bank Alpha has a negative gap corresponding to the [0–1M] bucket. This is mainly due to the magnitude of interest-bearing deposits. These instruments do not have any contractual maturity. They can be withdrawn at any moment and, theoretically, market risk changes may affect them instantaneously. However, Section 3.2.2 points out some sources of rigidity against interest rate fluctuations as well as dragging. Ideally, a market interest rate increase would have a negative impact on NII because of the first bucket and a positive one because of the (1M–3M], (3M–6M], and (6M–12M] buckets. No impact on the 1-year NII would derive from the (>12M) bucket and insensitive.

- **Margin at risk estimation.** Interest rate movements affect revenues as well as expenses. In particular, the following margin change (ΔNII) approximation holds[3]:

$$\Delta NII_h \approx \sum_{l=1}^{L \leq h} Gap_{rep,l} \cdot \overline{DD}_l \cdot f(\Delta R_l), \tag{3.7}$$

where $\overline{DD}_l$ is the period between l and the end of the holding period h (see Eq. 3.8) and $f(\Delta R_l)$ is a sensitivity function to risk-free movements. As a starting point, 100% sensitivity is usually applied. Nonetheless, Section 3.2.2 explores a more sophisticated approach by taking into account inflexibilities in the ongoing ALM. Finally, the following equation clarifies how to estimate

$$\overline{DD}_{A,l} = \frac{\sum_{i=1}^{N_l} A_{i,ra,t_{rep}\in l} DD_{i,t_{rep}\in l}}{\sum_{i=1}^{N_l} A_{i,ra,t_{rep}\in l}}, \tag{3.8}$$

where $DD_{i,t_{rep}\in l}$ represents the year fraction from the repricing time to the end of the holding period for all sensitive assets. The same mechanism applies to liabilities. Example 3.3 illustrates how to compute $\overline{DD}_{A,l}$ for Bank Alpha's (1M–3M] bucket.

Example 3.3 (1M–3M] Bank Alpha $\overline{DD}_{A,l}$ Computation

Let us consider Bank Alpha's (1M–3M] bucket ($A_{i,ra,t_{rep}\in l}$). In Table 3.8, the column headed $DD_{i,t_{rep}\in l}$ represents the time period (expressed in year fraction) between the repricing point and the end of the holding period. The next column $\overline{DD}_{A,l=(1M\text{-}3M]}$ shows how to compute Eq. (3.8).

---

3. A center dot is used in Eq. (3.7) to prevent misinterpretation.

Example 3.3 (1M–3M) Bank Alpha $\overline{DD}_{A,l}$ Computation—cont'd

TABLE 3.8 $\overline{DD}_{A,l}$ for Bank Alpha's (1M–3M) asset bucket (\$ billions)

		Assets		
		$A_{i,ra,t_{rep} \in l}$	$DD_{i,t_{rep} \in l}$	$\overline{DD}_{A,l}$
Cash resources				
	Cash			
	Interest-bearing deposits			
Securities				
	Trading account	1.00	0.90	
	Available for sale	4.00	0.80	
	Held to maturity			
Loans				
	Corporate loans	3.00	0.85	
	Retail loans			
	Real estate loans			
	Allowances for loan losses			
Other assets				
Total assets		8.00		0.83

As a final step, the margin at risk is estimated by application of an instantaneous interest rate shock (e.g., +1% and −1%; this corresponds to, say, $f(\Delta R_l) = \pm 1\%$).

In Bank Alpha's case, Table 3.9 highlights the key components to assess the margin at risk. The first row outlines each bucket's repricing gap. The second row shows the product $Gap_{rep,l} \cdot \overline{DD}_l$. The sum of these products is −\$27.42 billion. Consequently, a 1% increase in interest rates implies a −\$0.2742 billion impact on Bank Alpha's margin. In contrast, a 1% decrease in interest rates has a \$0.2742 billion impact on its margin. These variations account for 10.97% of the expected \$2.50 billion margin estimated over the period $[t_0, t_1]$ (see Table 3.2).

TABLE 3.9 Bank Alpha's Weighted Gap Analysis (\$ Billions)

	Gap				
	[0–1M]	(1M–3M)	(3M–6M)	(6M–12M)	Total
Gap	−59.50	8.00	21.00	14.99	
Weighted DD_t	0.92	0.83	0.70	0.40	
Gap · DD_t	−54.76	6.64	14.70	6.00	−27.42

M, Month(s).

Bank Alpha's margin at risk analysis emphasizes two main areas of additional investigation. On the one hand, its risk relies mainly on the [0–1M] bucket. As highlighted already, nonmaturing deposits play a key role in this gap, but they may be characterized by some inflexibility with respect to interest rate changes. On the other hand, the $\pm 1\%$ parallel shift of interest rates, usually applied in ALM analysis, is useful in terms of overall sensitivity analysis. However, for stress testing a more accurate term structure of interest rate analysis is required. The following two sections address these two topics.

Remark. It is worth noting that the symbol Δ is used to represent two concepts throughout the book. On the one hand, it denotes a variation (e.g., ΔNII). On the other hand, it refers to a macroeconomic scenario (i.e., $\mathbf{x}_\Delta$) when used as a subscript.

3.2.2 Interest Rate Sensitivity Analysis

Managing assets and liabilities is at the core of banking activity. Interest rates should align with market dynamics. However, bank rates are not always completely sensitive to market movements. This is specifically true when one is dealing with facilities without a contractual maturity (e.g., current accounts). Hence modeling the relationship between market and banking interest rates enhances the ALM process. Indeed, banks try to capture immediately and entirely any increase in market interest rates on their asset side. In contrast, they tend to postpone this increase on the liability side by dragging their preshock interest rates. The opposite occurs in the event of a market rate reduction.

A sensitivity analysis reveals how external shocks affect internal interest rates and eventually the NII. Kashyap et al. (2002) use a system of simultaneous equations to describe this relationship. Bellini and Riani (2010) propose an intuitive and easy-to-implement approach based on linear and nonlinear regression. They distinguish between elasticity and dragging as follows:

- **Elasticity.** This is estimated as a variation of banking interest rates in relation to a given change in market rates.
- **Dragging.** This quantifies the delay (dragging) in a bank interest rate adjustment against a market interest movement.

Let us start the explanation of this process by considering the relationship between a coupon rate, $C_{r,t}$ and the risk-free interest rate, R_t (e.g., LIBOR.[4]), at time $t = 1, \ldots, T$. The most intuitive way to capture the link between the bank coupon rate and the market rate is to use the following linear equation:

$$\Delta C_{r,t} = \vartheta_0 \Delta R_t + \vartheta_1 \Delta R_{t-1} + \cdots + \vartheta_k \Delta R_{t-k} + \epsilon_t, \tag{3.9}$$

where $\Delta C_{r,t}$ represents the first difference of $C_{r,t}$, while ΔR_{t-k} is kth difference of R_t. Errors (ϵ_t) are i.i.d. normally distributed.

---

4. For simplicity, hereafter this interest rate is considered as a risk-free proxy.

Eq. (3.9) measures the overall sensitivity to interest rate changes. However, it may be useful to disentangle elasticity and dragging contributions. According to Bellini and Riani (2010), the entire problem can be rewritten in terms of a constrained minimization. As detailed in Eq. (3.10), the objective is to minimize the square of the difference between $\Delta C_{r,t}$ (at time t) and a linear combination of risk-free interest changes ΔR at time $t, t-1, \ldots, t-k$. The minimization is performed by estimation of sensitivity ϑ and dragging $\eta = (\eta_1, \ldots, \eta_k)'$. The optimization is done by the imposition of some constraints. Firstly, the k dragging components work as weights in the adjustment process. Thus they need to sum up to unity. As a consequence, each dragging item is caused to vary between 0 and 1. The elasticity is constrained to be lower than 1 (in absolute value)[5] as detailed below:

$$\underset{\vartheta, \eta_0, \ldots, \eta_k}{\text{argmin}} \quad \sum_{t=k+1}^{T} \left[\Delta C_{r,t} - \vartheta (\eta_0 \Delta R_t + \eta_1 \Delta R_{t-1} + \cdots + \eta_k \Delta R_{t-k}) \right]^2$$

$$\text{s.t.}$$
$$\sum_{t=1}^{k} \eta_t = 1,$$
$$0 \leq \eta_t \leq 1,$$
$$|\vartheta| \leq 1.$$

(3.10)

Example 3.4 details how to estimate the parameters of Eq. (3.10).

Example 3.4 Interest Rate Elasticity and Dragging

Let us consider the quarterly loan interest rate time series from the first quarter of 2004 to the fourth quarter of 2014 for the Italian banking system. Interest rate changes correspond to $\Delta C_{r,t}$ in Eq. (3.10). The LIBOR acts as explanatory variable. The following MATLAB code serves the purpose of estimating elasticity and dragging parameters:

```
% 1. Load data
rawData = xlsread('Chap3rates.xlsx');
DeltaR = rawData(2:end,1) - rawData(1:end-1,1);
Loans = rawData(2:end,2);
DepositAccounts = rawData(2:end,3);
% 2. Define the Problem
k = 3;
% The number of lag terms / betas is k+1
whichCoupon = 'Loans';
% 3. Set up the predictors and responses
if strcmp(whichCoupon, 'Loans')
    C = Loans;
    else
    C = DepositAccounts;
end
Y = C(2:end) - C(1:end-1);
```

(Continued)

---

5. This relies on the assumption that a bank does not overreact to market movements.

Example 3.4 Interest Rate Elasticity and Dragging—cont'd

```
% And trim to accommodate k more lag terms:
Y = Y(1+k:end);
X = zeros(size(Y,1), k+1);
for iLag = 1:k+1
    X(:,iLag) = DeltaR(k-iLag+3:end - iLag +1);
end
obj = @(params) objectiveFcn(params, X, Y);
% 4. Run the optimization
param0 = [ones(k+1,1) / (k+1); 0];
% Constraint 1: All the betas sum to 1
Aeq = [ones(1, k+1), 0];
beq = 1;
% Constraints 2 and 3: betas lie between 0 and 1,
%and abs(theta) <= 1
lb = [zeros(k+1,1); -1];
ub = ones(k+2,1);
[paramEst,fval,exitflag] =...
fmincon(obj, param0, [], [], Aeq, beq, lb, ub);
eta = paramEst(1:end-1);
vartheta = paramEst(end);
%% Optimization function (used as a separate file)
function sse = objectiveFcn(params, X, Y);
vartheta = params(end);
eta = params(1:end-1);
Ypred = X * eta;
errors = Y - vartheta .* Ypred;
sse = errors' * errors;
% 5. Estimates
% eta =
% 0.5042
% 0.4875
% 0.0059
% 0.0024
% vartheta =
% 0.6525
```

According to the above, the overall sensitivity to LIBOR movements is $\vartheta = 0.6525$. Additionally, shocks are incorporated by banks with a 2-month dragging period (i.e., the sum of $\eta_1 = 0.5042$ and $\eta_2 = 0.4875$ accounts for 99.17% of the overall sensitivity).

Fig. 3.2 compares actual and fitted time series. R^2 is 66.98%.

Example 3.4 Interest Rate Elasticity and Dragging—cont'd

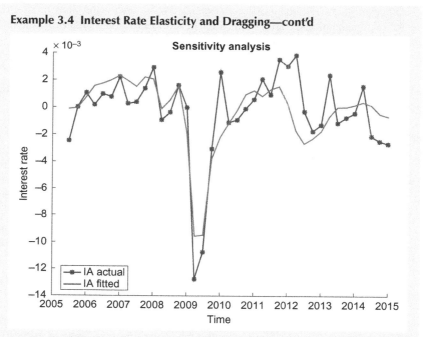

FIG. 3.2 Comparison of actual (first differences) and fitted Italian loan interest rate time series. *IA*, Asset interest.

In Section 3.2.1 a ±1% market interest rate variation was applied to assess Bank Alpha's margin at risk. Nonetheless, more realistic assumptions need to be followed for stress testing purposes. The stress testing scenario outlined in Chapter 2 described US short-term and long-term interest rate dynamics. Additional information on the entire term structure of interest rates is needed. In this sense, such a curve encompasses yields applied to different maturities. It enables investors to assess rates on short-term, medium-term, and long-term instruments. The next section describes how to estimate the term structure by use of affine models (Vasicek, 1977; Cox et al., 1985).

3.2.3 Term Structure of Interest Rates

A series of approaches have been developed to fit the term structure of interest rates. Diebold et al. (2006), among others, estimated a model that summarizes the yield curve using latent factors: level, slope, and curvature. They used a pure statistical approach enriched to include observable macroeconomic variables such as GDP, inflation, and the monetary policy to provide a characterization of the dynamic interactions between the macroeconomy and the yield curve.

Other authors followed a different perspective by anchoring their model to the following equation describing the present value of a default-free zero coupon bond with maturity τ:

$$V(\tau) = \mathbb{E}[e^{-\int_0^\tau R(u)du}], \tag{3.11}$$

where $R(\cdot)$ is the risk-free rate of interest, and the expectation is taken over all possible paths of $R(\cdot)$.

Vasicek (1977) proposed an affine (analytical) solution to the Eq. (3.11) as detailed below:

$$dR(t) = k[\theta - R(t)]dt + \sigma dW(t), \tag{3.12}$$

where $W(t)$ is a Wiener process, $\theta \geq 0$ is the long-run average interest rate, $k \geq 0$ is the mean reverting rate at which the process returns to its long-run mean, and $\sigma \geq 0$ is the volatility parameter of the process.

A few years later, Cox et al. (1985) extended this model by introducing a standard deviation factor $\sigma\sqrt{R(t)}$ that avoids the possibility of negative interest rates for all positive values of k and θ as detailed below:

$$dR(t) = k[\theta - R(t)]dt + \sigma\sqrt{R(t)}dW(t). \tag{3.13}$$

In the Cox-Ingersoll-Ross (CIR) model, when the rate is close to zero, the standard deviation also becomes very small, which reduces the effect of the random shock on the rate. Consequently, when the rate gets close to zero, its evolution becomes dominated by the drift factor, which pushes the rate upward. The term structure of interest rates may be estimated by use of the following equation:

$$R(\tau) = \frac{1}{\tau}[B(\cdot)R(t) - \ln(A(\cdot))], \tag{3.14}$$

where $A(\cdot)$ and $B(\cdot)$ contain the parameters θ, k, and σ and the risk premium parameter v, and τ denotes the maturity node of the term structure (Duan and Simonato, 1999).

Alternative techniques can be used to estimate $A(\cdot)$ and $B(\cdot)$. In what follows, the Kalman filter algorithm is used. In this regard the first step is to represent the above stochastic differential equations in a discrete time setting R_t.[6] Discrete time series are considered for each node (maturity date) of the term

---

6. In general, inference from discrete-time observations can be based on an approximation to the continuous-time likelihood, replacing Lebesgue integrals and Itô integrals by Riemann-Itô sums provided the observations times, as in our case, are closely spaced (see, e.g., Yoshida, 1992). On the other hand, if the time between observations is bounded away from 0, one should adjust the score function or, starting from the continuous-time likelihood function, should modify appropriately the estimating equations as suggested in Bibby and Sørensen (1995). Other alternative strategies are based on the Euler or Milstein scheme where, according to Bolder (2001), the stochastic differential equations can be solved for R_t and then this solution is discretized.

structure (e.g., nodes corresponding to 3 months, 12 months, and so on). Then a Kalman filtering process relies on a measurement and a transition equation as follows:

$$\mathbf{R}_t = \mathbf{Z}_t\boldsymbol{\alpha}_t - \mathbf{c}_t + \mathbf{G}\boldsymbol{\epsilon}_t, \tag{3.15}$$

$$\boldsymbol{\alpha}_{t+1} = \mathbf{T}_t\boldsymbol{\alpha}_t + \mathbf{d}_t + \mathbf{H}\boldsymbol{\epsilon}_t, \tag{3.16}$$

where Eq. (3.15) is the measurement equation, $\mathbf{R}_t$ is a vector containing the yield values for each node of the term structure, and $\mathbf{Z}_t$ is a matrix containing the coefficients $B(\cdot)$ in Eq. (3.14) to be applied to the vector state $\boldsymbol{\alpha}_t$. In this regard knowledge of the state allows us to predict the future dynamics of a deterministic system in the absence of noise. Further, $\mathbf{c}_t$ is a vector containing the coefficients $A(\cdot)$ in Eq. (3.14). Finally, $\mathbf{G}\boldsymbol{\epsilon}_t$ represents the error structure of the measurement equation. Eq. (3.16) is the so-called transition equation: $\boldsymbol{\alpha}_{t+1}$ is the one-step-ahead predicted state vector, $\mathbf{T}_t$ is a matrix of coefficients to be applied to the vector state $\boldsymbol{\alpha}_t$, $\mathbf{d}_t$ is another vector of coefficients, and $\mathbf{H}\boldsymbol{\epsilon}_t$ represents the error structure of the transition equation.

Appendix A enters into the details of the above Kalman filter. It shows how to obtain the equation to estimate the models of Vasicek (1977) and Cox et al. (1985). Moreover, Appendix B highlights how to detect atypical observations through the forward search analysis.

Example 3.5 shows how to estimate the term structure. It relies on the analytical representation in Appendix A to fit four nodes of the curve: 3 months, 3 years, 5 years, and 10 years.

Example 3.5 Estimation of the Term Structure of Interest Rates

The following daily time series from Jan. 2014 to Dec. 2014 feed the model by focusing on the following nodes: 3-month LIBOR, 3-year Interest Rate Swap, 5-year Interest Rate Swap, and 10-year Interest Rate Swap. The following MATLAB code estimates the term structure of interest rates by use of the CIR one-factor term structure detailed in Appendix A. It is worth mentioning that the inclusion of additional factors is straightforward, starting from the code listed below:

```
% Data upload
Y=xlsread('Chap3termstr2014');
global para0;
```

(Continued)

Example 3.5 Estimation of the Term Structure of Interest Rates—cont'd

```
% Set parameters
dt=1/250;
ratestart=mean(Y(end,1));
ttime=250;
nrow=ttime;
tau=[0.33,3,5,10];
ncol=size(tau,2);
Delta=[1,1,1,-1,ones(1,numel(tau))]';
%starting values
para0=[0.04,0.8,0.01,-0.002,1e-2*rand(1,ncol).*ones(1,ncol)];
nparam=size(para0,2);
para_BFGS=log(para0./(Delta'-para0));
options = optimset('Display','off','MaxIter',1000,...
'MaxFunEvals',1e2,'TolX',1e-4,'TolFun',10^(-8),...
'HessUpdate', 'bfgs');
% Maximization procedure
% Functions LLoneCIR_BFGS and LLoneCIR_param_BFGS required
for imax=1:3
    x='';
    try
        [x]=fminunc(@LLoneCIR_BFGS,para_BFGS,...
            options,Y, tau, nrow, ncol, Delta,dt);
    catch exception
    end
    if ~isempty(x) || imax==3
        break;
    else
        % restart maximization with a new set of starting
values
        para1=para0+1e-4*randn(1,length(para0));
        para_BFGS=log(para1./(Delta'-para1));
    end
end
if imax==3;
    x=para_BFGS;
end
[Theta]=LLoneCIR_param_BFGS(x,Y, tau,...
Delta, dt, nrow, ncol,nparam);
% Theta' =
% 0.0400 0.8001 0.0101 -0.0019 0.0064 0.0009 0.0029 0.0056
```

Fig. 3.3 outlines the shape of the term structure of interest rates estimated above.

Example 3.5 Estimation of the Term Structure of Interest Rates—cont'd

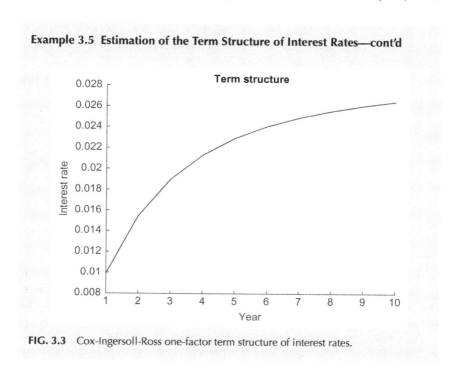

FIG. 3.3 Cox-Ingersoll-Ross one-factor term structure of interest rates.

The next section embraces all modeling components above described in order to assess the margin at risk under a specific stress testing scenario.

3.2.4 Margin at Risk Under a Stress Testing Scenario

In this step of the process, a quick summary of the entire process may be helpful to recap all the concepts introduced so far. Firstly, an asset and liability framework was presented. Then a profit and loss analysis focused on the mechanism to compute the NII (i.e., interest margin). A distinction was made between a rolling and a growing balance sheet perspective. A study of asset and liability gaps allowed us to map financial instruments among repricing buckets. The margin at risk ended up as the product of a shocked interest rate and gaps (weighted according to the remaining term at the end of the holding period h). A sensitivity inspection enriched the research by illustrating the difference between elasticity and dragging contributions. Finally, a study of the term structure of interest rates aimed to remove the usual $\pm 1\%$ shift used for ALM enquiries. The next step is to analyze the margin at risk due to a specific scenario.

Knowledge of the term structure curve facilitates the NII analysis under a macroeconomic stress path. Indeed, the vector of stressed macroeconomic variables $\mathbf{x}_\Delta = (x_{1,\Delta}, \ldots, x_{p,\Delta})'$ drives interest rate fluctuations. Therefore a coherent transmission process is vital. More specifically, a coupon rate is a function of $\mathbf{x}_t$ by means of the risk-free interest rate $C_t[R_t(\mathbf{x}_t)]$. Therefore the coupon rate under stress may be approximated as follows:

$$C_{t,\Delta} \approx C_t[R_t(\mathbf{x}_{t_0})] + g[\Delta R_t(\mathbf{x}_\Delta)], \tag{3.17}$$

where g is interpretable as Eq. (3.10), while $\Delta R_t(\mathbf{x}_\Delta)$ denotes the coupon variation due to $\mathbf{x}_\Delta$. This process allows us to estimate the shocked NII as follows:

$$\Delta NII_{h,\Delta} \approx \sum_{l=1}^{L} Gap_{rep,l} \cdot \overline{DD}_l \cdot g[\Delta R_l(\mathbf{x}_\Delta)]. \tag{3.18}$$

The next section details how to apply the above mechanism to Bank Alpha.

3.2.5 Bank Alpha's Stress Testing Margin at Risk

As the final step of the margin at risk analysis, Bank Alpha allows us to highlight the impact of a given scenario in a bank. For clarity, a rolling scheme applies. This assumption will be removed in Chapter 5, where balance growth will impact the stress testing analysis.

As a starting point, the t_0 term structure is assumed to be the curve estimated in Example 3.5. The macroeconomic scenario used for Bank Alpha's stress testing is applied as a parameter constraint to estimate the stressed interest rate curve displayed in Fig. 3.4. On this subject, Exercise 3.2 details how to impose

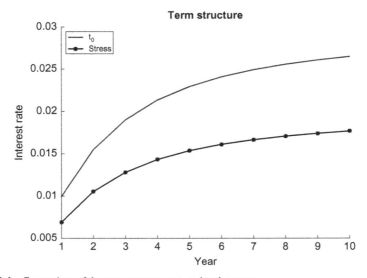

FIG. 3.4 Comparison of the term structure at t_0 and under stress.

constraints in a Monte Carlo simulation process. It serves the purpose of both stress testing and risk integration.

The margin at risk is a function of the difference between the term structure at t_0 and the stressed one.[7] Table 3.10 highlights that assets and liabilities are aggregated according to their corresponding pivotal interest rate (i.e., the interest rate on which the repricing is based) and repricing buckets. The sensitivity column highlights the proportion of interest rate shock captured by Bank Alpha (this sensitivity takes into account the multicountry nature of the bank). Then the rate change graphically outlined in Fig. 3.4 is applied to each pivotal interest rate. Finally, Eq. (3.18) approximates the NII variation due to interest rate movements. As a result, Bank Alpha's NII reduces by $0.10 billion over a 1-year horizon. This result will be compared with the NII variation estimated in Chapter 5 by the consideration of a dynamic balance sheet. Meanwhile, the reader is invited to solve Exercise 3.3 to grasp the key ideas discussed above.

Changes in interest rates affect not only the margin but also the present value of both assets and liabilities. The next section describes how a bank measures this risk. Specific attention is devoted to the trading book VaR.

3.3 VALUE AT RISK

The previous sections focused on margin changes due to economic movements over a given holding period h (e.g., 1 year). However, asset and liability values also incorporate macroeconomic shocks.

Notwithstanding potential interest in a full fair value assessment extended to all assets and liabilities, regulatory requirements and accounting standards rely on a more conservative historical or amortizing cost perspective for the vast majority of asset and liability instruments. Nevertheless, it is worth anticipating the present value estimation mechanism used in Chapter 7 for risk integration purposes. This framework will be used in Sections 3.3.1–3.3.5 on a more restricted balance sheet area (i.e., trading book portfolio) to assess the VaR.

If PV denotes the present value for a zero coupon bond asset $A_{i,T}$, the following equation holds:

$$PV(A_{i,T}, R_T(\mathbf{x}_0)) = A_{i,T} e^{-R_T(\mathbf{x}_0) \cdot T}, \tag{3.19}$$

where T is the maturity of the bond, while $R_T(\mathbf{x}_0)$ identifies the risk-free interest rate at time T, which is a function of the ongoing ($t = 0$) vector of macroeconomic variables $\mathbf{x}_0$ (Duffie and Singleton, 2003). Hence the present value of asset A_i can be estimated as follows:

---

7. On this subject, Bank Alpha is analyzed at a consolidated level. Each country's adjustment is captured via interest rate sensitivity.

TABLE 3.10 Bank Alpha's Margin at Risk (Net Interest Income Variation) Under Stress Including Elasticity ($ Billions)

Pivotal interest rate	Assets						Total	Sensitivity	ΔR	ΔNII
	[0–1M]	(1M–3M]	(3M–6M]	(6M–12M]	(>12M]	Insensitive				
3 months	1.00			6.00	6.00		13.00	95.00%	−0.30%	−0.01
6 months	2.00		10.00	3.00	6.25		21.25	95.00%	−0.30%	−0.03
12 months			12.50	0.60	8.86		21.96	90.00%	−0.30%	−0.02
3 years		1.00		4.00	2.40		7.40	90.00%	−0.62%	−0.02
5 years		4.00		6.25			10.25	90.00%	−0.75%	−0.05
7 years		3.00		11.14			14.14	82.00%	−0.82%	−0.06
≥10 years			1.00				1.00	82.00%	−0.88%	−0.00
Total assets	3.00	8.00	23.50	30.99	23.51	11.00	100.00			−0.20

TABLE 3.10 Bank Alpha's Margin at Risk (Net Interest Income Variation) Under Stress Including Elasticity ($ Billions)—cont'd

| Pivotal interest rate | Liabilities | | | | | | Total | Sensitivity | ΔR | ΔNII |
	[0–1M]	(1M–3M]	(3M–6M]	(6M–12M]	(>12M]	Insensitive				
3 months	62.50		2.50	3.00			68.00	45.00%	−0.30%	0.08
6 months				6.00			6.00	50.00%	−0.30%	0.01
12 months				3.00			3.00	55.00%	−0.30%	0.00
3 years				4.00	5.00		9.00	65.00%	−0.62%	0.01
5 years										
7 years										
≥10 years										
Total liabilities	62.50		2.50	16.00	5.00	14.00	100.00			0.10
ΔNII										−0.10

M, Month(s).

$$PVA_i = \sum_{t=0}^{T_i} PV(A_{i,cf,t}, R_t(\mathbf{x}_0)), \tag{3.20}$$

where $PV(\cdot)$ is the function described in Eq. (3.19) applied to each cash flow $A_{i,cf,t}$ (i.e., the subscript cf stands for cash flow). In this regard, it is worth highlighting that in Section 3.2 on the margin at risk analysis, the focus was on the residual amount $A_{i,ra,t}$ at each repricing time t. In contrast, the present value estimation focuses on cash flows. The present value of A_i estimated according to $\mathbf{x}_\Delta$ is as follows:

$$PVA_{i,\Delta R} = \sum_{t=0}^{T_i} PV(A_{i,cf,t}, R_t(\mathbf{x}_\Delta)). \tag{3.21}$$

The following equation encompasses the value change for all assets due to interest rate variations:

$$\Delta A_{\Delta R} = \sum_{i=1}^{N} (PVA_{i,\Delta R} - PVA_i). \tag{3.22}$$

Chapter 7 introduces additional risk sources (e.g., credit risk) to obtain a full fair value estimate. However, hereafter only interest rate fluctuations are taken into account.

For each liability L_j, the following equation holds:

$$PVL_j = \sum_{t=0}^{T_j} PV(L_{j,cf,t}, R_t(\mathbf{x}_0)), \tag{3.23}$$

where $L_{j,cf,t}$ stands for liability L_j cash flow at time t. As a consequence, the liability present value estimated according to $\mathbf{x}_\Delta$ is as follows:

$$PVL_{j,\Delta R} = \sum_{t=0}^{T_j} PV(L_{j,cf,t}, R_t(\mathbf{x}_\Delta)). \tag{3.24}$$

The total value change of a liabilities due to interest rate variations is

$$\Delta L_{\Delta R} = \sum_{j=1}^{M} (PVL_{j,\Delta R} - PVL_j). \tag{3.25}$$

An overall bank loss distribution relies on different interest rate scenarios (i.e., set of rate curves $R_{t,\Delta}(\mathbf{x}_\Delta)$). For each of them the following loss holds:

$$Loss_{AL,\Delta R} = \Delta A_{\Delta R} - \Delta L_{\Delta R}, \tag{3.26}$$

where $Loss_{AL,\Delta R}$ is the asset and liability loss due to interest rate changes ΔR.

A given quantile of the loss distribution synthetically measures the overall bank VaR due to interest rate movements. The VaR at the $(1 - \alpha)$ level is computed as the smallest loss (ℓ) such that

$$VaR_{AL,(1-\alpha)} = \inf_\ell P(Loss_{AL,\Delta R} > \ell) \leq \alpha. \tag{3.27}$$

Acerbi and Tasche (2002) showed that the VaR suffers from two severe deficiencies if it is considered as a measure of downside risk. In fact, it is not subadditive and it is insensitive to the size of the loss beyond the prespecified threshold. Subsequently, an additional risk measure is the expected shortfall (ES), which corresponds to the expected loss exceeding the VaR

$$ES_{AL,(1-\alpha)} = E\left(Loss_{AL,\Delta R}|Loss_{AL,\Delta R} > VaR_{AL,(1-\alpha)}\right) = \frac{1}{\alpha}\int_{u>1-\alpha}^{1} VaR_{AL}(u)du. \tag{3.28}$$

The ES is subadditive, and it provides information on the amount of loss exceeding the VaR. Portfolios with a low ES should also have a low VaR. Additionally, under general conditions, the ES is a convex function and it is a coherent measure of risk as well.

Despite the mathematical disadvantages of the VaR, this is the measure currently used in practice to assess market risk for both managerial and regulatory purposes. According to what was stated above, in practice a fair value estimation applies only to the trading book. A full fair value assessment will be investigated in Chapter 7 as a crucial engine of the risk integration process. In what follows, attention is devoted to the three main techniques to assess market VaR: variance-covariance, Monte Carlo simulation and historical simulation. The next section describes the variance-covariance method, while later sections will outline Monte Carlo and historical simulation VaR.

3.3.1 Variance-Covariance Value at Risk

According to the variance-covariance approach, the return of each instrument belonging to the trading book portfolio is assumed to follow a normal distribution $N(\mu, \sigma^2)$. A series of studies show that this assumption hardly holds in practice (Christoffersen, 2012). Nonetheless, it can be considered as the first step to develop a more comprehensive framework.

Let us consider the VaR of a portfolio over a 1-day horizon with a 5% probability. The VaR is computed on the basis of the following steps:

- **Portfolio value at t_0.** A mark-to-market or mark-to-model estimation of the portfolio is performed to assess its value P_0.
- **Portfolio value after 1 day.** The portfolio value after 1 day is a function of P_0 and a 1-day horizon return r as follows: $P_1 = f(P_0, e^r)$.
- **Portfolio loss over 1-day.** Losses are estimated on the basis of $\hat{P}_1 - P_0$.

It is worth remarking that the VaR can be written as $P_0(1 - e^{\hat{r}})$. Hence, for $\hat{r}$ sufficiently small, $e^{\hat{r}} \approx 1 + \hat{r}$, VaR can be approximated through $P_0 \hat{r}$. Therefore the key element to be modeled is the forecast return $\hat{r}$.

Example 3.6 highlights the key features of the variance-covariance VaR.

Example 3.6 Variance-Covariance VaR for a Single Financial Instrument

A financial instrument has a current value of $1 million. Under the normality assumption, its return is estimated as follows:

$$\hat{r} = \Phi(\alpha)^{-1}\sigma_{1|0} + \mu_{1|0},$$

where $\Phi(\alpha)^{-1}$ is the inverse normal cumulative distribution function.

Firstly, the expected portfolio return $\mu_{1|0}$ is foreseen by our aiming to compute the 1-day VaR. A zero mean return is assumed over a 1-day horizon. Thus, considering $\sigma_{1|0} = 0.04$, the worst return, at a confidence level $\alpha = 5\%$, is estimated as

$$\hat{r} = -1.65\sigma_{1|0} + \mu_{1|0},$$

which, in the above case, corresponds to $\hat{r} = -1.65 \cdot 0.04 + 0 = -0.066$. Hence the value of the financial instrument is $1 \cdot e^{-0.066} = \$0.936$ million.

The VaR at the 5% confidence level over the 1-day horizon is $0.064 million.

In a portfolio composed of financial instruments having linear payoffs, the return has variance $\mathbf{w}'\Sigma\mathbf{w}$, where $\mathbf{w}$ is the vector of weights and Σ is the variance-covariance matrix of returns. The following equation summarizes the variance of a portfolio as a linear sum of its assets: $\sigma_w^2 = \mathbf{w}'\Sigma\mathbf{w}$, where the subscript w highlights the weight structure of the portfolio. The usual assumption is to multiply the 1-day VaR by the square root of t. Therefore the following holds:

$$VaR_{(1-\alpha),t} = \Phi^{-1}(\alpha)\sqrt{\mathbf{w}'\Sigma\mathbf{w}}\sqrt{t}. \tag{3.29}$$

This process is shown in Example 3.7 for a two-instrument portfolio.

Example 3.7 Variance-Covariance VaR for a Two-Instrument Portfolio

Let us assume a portfolio is made up of asset A worth $1 million and asset B worth $3 million. Asset A has a daily variance of 0.2%, while asset B has a daily variance of 0.3%. Their covariance is 0.06%. The portfolio variance is as follows:

$$\mathbf{w}'\Sigma\mathbf{w} = \begin{pmatrix} 1.00 & 3.00 \end{pmatrix} \begin{pmatrix} 0.0020 & 0.0006 \\ 0.0006 & 0.0030 \end{pmatrix} \begin{pmatrix} 1.00 \\ 3.00 \end{pmatrix} = 0.0326.$$

In line with Eq. (3.29), the following holds:

$$VaR_{95\%} = 1.65 \cdot \sqrt{0.0326} \cdot \sqrt{10} = \$0.942 \text{ million}.$$

Real portfolios are usually too large for an asset-level representation. Hence a portfolio is mapped on m risk factors, and its variance assumes the following shape: $\mathbf{w}'\mathbf{B}\Psi\mathbf{B}'\mathbf{w}$. In this regard, $\mathbf{w}$ is the $n \times 1$ vector of each financial instrument portfolio weight, $\mathbf{B}$ is the $n \times m$ matrix of factor sensitivities, and Ψ is the $m \times m$ matrix of factor variance-covariance. The VaR is computed as follows:

$$VaR_{(1-\alpha),t} = \Phi^{-1}(\alpha)\sqrt{\mathbf{w}'\mathbf{B}\Psi\mathbf{B}'\mathbf{w}}\sqrt{t}. \tag{3.30}$$

For portfolios containing options, the above representation is no longer accurate. A delta approximation for complex instruments is very convenient. It allows us to capture variations due to changes in the underlying asset price. However, when one is considering noninfinitesimal changes, high-order moments are required as described, among others, by McNeil et al. (2015). For a comprehensive discussion, the reader may refer to Christoffersen (2012).

The following section faces the issue to estimate the VaR in a Monte Carlo simulation setting.

3.3.2 Monte Carlo Simulation Value at Risk

The idea behind the variance-covariance VaR can be enriched and used to extend the VaR computation to a wider range of financial instruments by the simulation of their movements or the dynamics of the underlying factors.

Scenario generations need to be coherent with potential portfolio losses. Hence the variance of each instrument's return as well as the covariance among instruments is considered. A Monte Carlo process generates financial returns. When a normal distribution is assumed, the process is summarized through the following three main steps:

1. **Cholesky decomposition.** The return's covariance matrix is factorized into the product of two matrices $\Sigma = \mathbf{Q}\mathbf{Q}'$ according to a Cholesky decomposition.
2. **Simulation.** Random realizations are generated from a standard normal variable (i.e., $N(0, 1)$).
3. **Covariance structure.** The above random generations are premultiplied by the matrix $\mathbf{Q}$ to convert an i.i.d. $N(0, 1)$ sample into simulated returns with an appropriate covariance structure.

Example 3.8 highlights how to implement the three-step simulation algorithm described above.

Example 3.8 Monte Carlo Return Simulation

Let us consider a three-asset portfolio with a return variance-covariance structure defined as follows:

$$\Sigma = \begin{pmatrix} 2.00 & 0.70 & 0.60 \\ 0.70 & 1.30 & 0.30 \\ 0.60 & 0.30 & 1.70 \end{pmatrix}.$$

The following MATLAB code exemplifies how to obtain 100 simulated returns by use of the process described above:

```
%Number of assets
nasset=3;
%Number of simulations
nsim=100;
%Correlation matrix
Rmat=
[2.00 0.70 0.60
0.70 1.30 0.30
0.60 0.30 1.70];
% 1. Cholesky decomposition of Rmat
Q=chol(Rmat);
% 2. Simulation of iid from N(0,1)
zz = normrnd(0,1, nasset,nsim);
% 3. Simulation of returns
rr=Q*zz;
```

The portfolio value distribution derives from the application of financial instruments' pricing models to simulated returns. Subsequently, the portfolio VaR is obtained as the percentile of the overall loss distribution.

The next section provides a brief description of historical simulation VaR.

3.3.3 Historical Simulation Value at Risk

The historical simulation VaR originates from the generation of the empirical density of the portfolio's returns. Assumptions are required neither for the distribution of returns nor for their comovements.

In a linear portfolio the return, $r_{w,t}$ can be represented as a weighted sum of the returns of each asset $k = 1, \ldots, K$ as follows:

$$r_{w,t} = w_1 r_{1,t} + \cdots + w_K r_{K,t}. \tag{3.31}$$

This representation allows the simulation of h-day-ahead portfolio profit and loss based on historical returns $r_1, \ldots, r_K$.

A full evaluation of all instruments would be desirable when the portfolio includes options and other complex products. Nevertheless, an approximation based on delta-gamma or delta-gamma-vega is often used in practice (Christoffersen, 2012). In this regard, firstly current portfolio deltas, gammas, and vegas are computed. Then they are applied to the historical series on the underlying assets and risk factors.

The next section introduces the key differences between the VaR and the stressed VaR (SVaR) as well as their use for regulatory purposes.

3.3.4 Stress Testing and Regulatory Value at Risk

The SVaR relies on volatilities and correlations experienced during a historical adverse event or representative of an extreme scenario. When the variance-covariance approach is used, extreme market conditions are taken into account by replacement of the current variance-covariance matrix with the stressed one. Also in the case of historical VaR, the stress test can be run by use of a stressed variance-covariance matrix. Let us consider the vector of returns $\mathbf{r}$, the current covariance matrix $\mathbf{\Sigma}$, and the covariance under stress $\mathbf{W}$. Stressed returns are simulated by use of the Cholesky decomposition of $\mathbf{\Sigma}$ and $\mathbf{W}$ denoted respectively as $\mathbf{C}$ and $\mathbf{D}$ as follows:

$$\mathbf{r}_{stress} = \mathbf{CD}^{-1}\mathbf{r}. \tag{3.32}$$

In the case of a breakdown of volatilities and correlations, a variance-covariance decomposition may be very useful to shock volatilities separately from correlations. This decomposition can be represented as $\mathbf{V} = \mathbf{AHA}$, where $\mathbf{H}$ is the correlation matrix and $\mathbf{A}$ is the diagonal matrix with standard deviations along the diagonal.

From a regulatory standpoint, the market risk is measured as the sum of the following components: regulatory VaR, regulatory SVaR, incremental risk charge (IRC), and standard specific risk charge (SSRC) as detailed below:

- **Regulatory VaR.** Under Basel III (BIS, 2011), the regulatory VaR is the estimate of the potential decline in the value of a position or a portfolio under normal market conditions. The regulatory VaR uses a 10-day horizon and is calibrated to a $(1 - \alpha) = 99\%$ confidence level. For regulatory purposes, the following equation holds:

$$VaR_{reg,t} = \max\left(VaR_{(1-\alpha),t}, \frac{1}{60}\sum_{r=1}^{60} VaR_{(1-\alpha),t-r} \cdot MM\right), \tag{3.33}$$

where $VaR_{(1-\alpha),t}$ is the end-period 10-day VaR. The other item in the *max* formula (i.e., $\frac{1}{60}\sum_{r=1}^{60} VaR_{(1-\alpha),t-r} \cdot MM$) is the average 10-day VaR measured over the 60 days before the end of the period. *MM* is a multiplier ranging between 3 and 4, based on the number of back-testing exceptions that occur in a rolling 12-month period. On this subject, banks are required to perform back-testing to evaluate the effectiveness of their models. This is a process through which the 1-day VaR, at the 99% confidence level, is compared with the buy-and-hold daily revenue (i.e., the profit and loss impact if the portfolio is held constant at the end of the day and repriced the following day).

- **Regulatory SVaR.** The regulatory VaR models are usually used to assess the SVaR. The main difference between VaR and SVaR is in the parameters used for their estimate. More specifically, regulatory SVaR is based on model parameters (such as volatilities and correlations) calibrated on historical data from a continuous 12-month period reflecting significant financial stress. The regulatory SVaR is periodically calibrated by means of internal methods and policies to determine the severest stress period for a bank's current trading book.

- **IRC.** The aim of the IRC is to cover default and credit migration risks. The IRC is measured over a 1-year horizon at the 99.9% confidence level under the assumption of constant positions. Liquidity horizons establish the effective holding period of the assets and are defined as the time that would be required to reduce exposure, or hedge all material risks, in a stressed market environment. A constant position assumption means that a bank maintains the same set of positions throughout the 1-year horizon (regardless of the maturity date of the positions) so as to model profit and loss distributions. IRC models are usually designed to capture market-specific and issuer-specific concentrations, credit quality, and liquidity horizons. They also aim to recognize the impact of correlations between default and migration events among issuers.

- **SSRC.** Specific risk is due to loss from changes in the market value of a position that could result from factors other than market movements and includes event risk, default risk, and idiosyncratic risk. SSRCs include any debt or equity position that has not received a modeled-specific risk charge (i.e., regulatory VaR, SVaR, or IRC). On the basis of the Basel II Accord rules, one usually derives SSRCs by taking a percentage of the market value. This percentage depends on the product type, the time to maturity, the bank's internal credit rating, and other factors.

According to the above, the RWA due to market risk is computed as follows:

$$RWA_{mkt} = 12.5(VaR_{reg} + SVaR_{reg} + IRC + SSRC), \qquad (3.34)$$

where 12.5 is the reciprocal of 8%, corresponding to the minimum capital requirement.

The next section uses Bank Alpha to show how to compute market risk RWA.

TABLE 3.11 Bank Alpha's Value at Risk and Stressed Value at Risk ($ Billions)

	Portfolio Value	1-day VaR	10-day VaR	1-day SVaR	10-day SVaR
Equities	0.80	0.0017	0.0054	0.0145	0.0459
Bonds	1.70	0.0013	0.0041	0.0116	0.0367
Others	0.50	0.0002	0.0006	0.0014	0.0044
Total	3.00	0.0032	0.0101	0.0275	0.0870

SVaR, Stressed value at risk; VaR, value at risk.

3.3.5   Bank Alpha's Market RWA

In what follows, Bank Alpha's trading portfolio is analyzed by our distinguishing among equities, bonds, and other financial instruments. A Monte Carlo simulation approach is used to compute both VaR and SVaR. Table 3.11 points out a $0.0101 billion 10-day VaR, which accounts for 0.34% of the outstanding value. On the other hand, the 10-day SVaR is $0.087 billion, which represents 2.90% of the trading account value.

With regard to the regulatory perspective, Table 3.11 shows that the 10-day VaR is $0.0101 billion. In contrast, the 60-day average 10-day VaR is $0.0080 billion. On the latter, a $MM = 3$ regulatory multiplier is applied, leading to $0.0240 billion. According to Eq. (3.33), the following holds: $VaR_{reg} = \max(0.0101, 0.0240) = \$0.0240$ billion.

A similar approach holds for $SVaR_{reg}$. In this case the 10-day SVaR is $0.0870 billion, while the 60-day SVaR multiplied by the factor MM is $0.0904 billion. As a result, $SVaR_{reg} = \max(0.0870, 0.0904) = \$0.0904$ billion.

In line with Eq. (3.34), two additional components are required. On the one hand, Bank Alpha's IRC is $0.0600 billion. On the other hand, the SSRC accounts for $0.0104 billion. Table 3.12 summarizes Bank Alpha's market risk capital charge and RWAs.

The next section uses the framework outlined in Section 3.2 to describe how to assess the liquidity risk.

3.4   LIQUIDITY ANALYSIS

As shown in Chapter 1, liquidity issues can be crucial for bank solvency. New requirements were introduced after the 2007–09 crisis by strengthening the way banks manage their cash flows. The focus of this section is on the asset and liability mechanism used for liquidity purposes. This analysis aligns with managerial practice and is a crucial tool for the treasury department. In contrast, Chapter 6 will enter into the details of the regulatory liquidity ratios: liquidity

TABLE 3.12 Bank Alpha's Market Risk Capital Charge and Risk-Weighted Assets ($ Billions)

	CC	RWAs
VaR	0.0240	0.30
SVaR	0.0904	1.13
IRC	0.0600	0.75
SSRC	0.0104	0.13
Total	0.1848	2.31

IRC, Incremental risk charge; CC, capital charge; RWAs, risk-weighted assets; SSRC, standard specific risk charge; SVaR, stressed value at risk; VaR, value at risk.

coverage ratio and net stable funding ratio. Hence the reader is invited to study this section in conjunction with the regulatory discussion provided later in the book.

In line with an ALM perspective, cash inflows and outflows are aggregated over subinterval grids as per the margin at risk analysis described in Section 3.2.1. The key difference relies on the trigger to group financial instruments around time buckets. If the margin at risk relied on the repricing schedule, the liquidity analysis would be based on the cash flow timeline as listed below:

- **Split between contractual and noncontractual maturity cash flows.** Firstly, operations with a contractual cash flow structure are separated from other operations. For example, current accounts do not have a contractual maturity.
- **Time grid bucketing.** The second step is to define cash flow buckets on which to root the aggregation process (e.g., [0–1M], (1M–3M]).
- **Gap analysis.** As per the margin at risk, a mapping process allows us to allocate cash flows to time buckets. Gaps are computed as follows:

$$Gap_{liq,ll} = \sum_{i=1}^{N_{ll}} \mathbb{1}_{i,liq,t_{liq}\in ll}A_{i,cf,t_{liq}} - \sum_{j=1}^{M_{ll}} \mathbb{1}_{j,liq,t_{liq}\in ll}L_{j,cf,t_{liq}}, \qquad (3.35)$$

where $\mathbb{1}_{i,liq,t_{liq}\in ll}$ is 1 when a cash flow occurs within the bucket ll,

$$\mathbb{1}_{i,liq,t\in ll} = \begin{cases} 1 & \text{for} \quad t_{liq} \in ll, \\ 0 & \text{otherwise.} \end{cases}$$

$A_{i,cf,t_{liq}\in ll}$ represents the cash flow (indexed by cf) of customer i within the bucket ll. In contrast, in margin at risk analysis the focus was on repricing (i.e., a financial instrument was allocated to the repricing bucket). A similar notation applies to liabilities.

Example 3.9 shows how liquidity gaps are computed for the one-asset, one-liability portfolio introduced in Example 3.2.

Example 3.9 Liquidity Analysis for a One-Asset, One-Liability Portfolio

Let us consider the portfolio described in Example 3.2. Fig. 3.5 outlines the cash flow structure for both assets and liabilities. For simplicity, interest flows are excluded from the analysis.

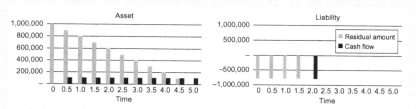

FIG. 3.5 Cash flow and residual amount ($).

The first bucket covers the first year [0–1Y] (see Table 3.13). The other two buckets follow a similar structure: (1Y–3Y), (3Y–6Y]. Cash inflows of $0.10 million occur each at the end of Dec. and end of Jun. An overall cash outflow occurs at the end of Dec. in 2 years.

TABLE 3.13 Liquidity Gap Analysis for the One-Asset, One-Liability Portfolio of Example 3.2

	[0–1Y]	(1Y–3Y]	(3Y–6Y]
Assets	0.2	0.4	0.4
Liabilities		−0.8	
Gap	0.2	−0.4	0.4

Y, Year(s).

The next section focuses on Bank Alpha's liquidity.

3.4.1 Bank Alpha's Liquidity Analysis

Table 3.14 summarizes Bank Alpha's main liquidity features. It highlights that the bank has very few exposures with long-term maturity. Hence the liquidity gap concentrates mainly on the short period.

One needs to bear in mind that deposits (interest bearing) include nonmaturity instruments in the [0–1M] bucket. From a practical point of view, these

TABLE 3.14 Bank Alpha's Liquidity Analysis ($ Billions)

		Assets								
		[0–1M]	[1M–3M]	[3M–6M]	[6M–12M]	[1Y–2Y]	[2Y–5Y]	[5Y–10Y]	(>10Y)	Total
Cash resources	Cash									6.00
Securities	Interest-bearing deposits				6.00					13.00
	Trading account	1.00	1.00							
	Available for sale		1.00				6.00			
	Held to maturity					3.00	1.00			
Loans	Corporate loans	2.00	1.00	2.00	5.00	5.00	10.00			70.00
	Retail loans			2.00	3.00	10.00	10.00			
	Real estate loans				1.00	3.00	7.00	4.00	5.00	
Other assets										
Total assets		3.00	3.00	4.00	15.00	21.00	34.00	4.00	5.00	89.00

Liabilities

	[0–1M]	(1M–3M)	(3M–6M)	(6M–12M)	(1Y–2Y)	(2Y–5Y)	(5Y–10Y)	(>10Y)	Total
Deposits									70.00
Non–interest bearing		5.00	2.5						
Interest bearing	62.50								
Other liabilities									17.00
Acceptances				3.00					
Bonds				1.00	5.00	5.00			
Other residual liabilities					3.00				
Subordinated debts				1.00		3.00			4.00
Noncontrolling interests									
Shareholder equity									
Common shares									
Preferred shares									
Retained earnings									
Total liabilities	62.50	5.00	2.5	5.00	8.00	8.00			91.00
Gap	–59.50	–2.00	1.50	10.00	13.00	26.00	4.00	5.00	–2.00

M, Month(s); Y, year(s).

liabilities are usually stable but they may become volatile during liquidity crises (see, e.g., the Cyprus Popular Bank Business Case 8.2 in Chapter 8).

Chapter 6 devotes specific attention to nonmaturity instruments while studying regulatory liquidity and net stable funding ratios. Additionally, Chapter 7 scrutinizes the peculiarities of nonmaturity deposits from a risk integration point of view.

3.5 SUMMARY

An ALM framework was introduced by our firstly focusing on the margin at risk analysis. A formal modeling was accompanied by illustrative examples showing how to map financial instruments according to their repricing structure. Bank Alpha illustrated the key steps of the asset and liability process by our paying specific attention to interest rate sensitivity analysis (i.e., elasticity and dragging components). Given the importance of interest rate shocks in assessing the margin at risk, the term structure of interest rates was explored. An affine representation due to Vasicek (1977) and Cox et al. (1985) was investigated and fitted via Kalman filtering. This study served as a reference to assess Bank Alpha's 1-year margin at risk based on a rolling balance sheet scheme.

Given that macroeconomic changes affect not only a bank's interest margins but also the present value of assets and liabilities, a VaR framework was proposed. Three methods were described with regard to the trading book: variance-covariance, Monte Carlo, and historical simulation. A distinction was made between managerial and regulatory measures. Furthermore, the SVaR entered the scope of the study.

Finally, the last part of the chapter focused on liquidity analysis. An asset and liability approach was described and its logic was applied to Bank Alpha.

SUGGESTIONS FOR FURTHER READING

ALM in banks is explored in many books, among which are Choudhry (2007) and Bohn and Elkenbracht-Huizing (2014). The development of risk management practice has been accompanied by a series of interesting texts on VaR and financial instrument pricing. The reader may certainly benefit from the introduction to options, futures and other derivatives presented by Hull (2012). A more advanced mathematical description is given by Brigo and Mercurio (2006), while Christoffersen (2012) balances a rigorous statistical formalization with a large number of examples and applications.

The term structure of interest rates is studied in a series of articles, among which that by Diebold et al. (2006) focuses on a statistical fitting approach by linking level, slope, and curvature to economic variables. Vasicek (1977) and Cox et al. (1985) pioneer a wide literature on interest rate modeling. Duan and Simonato (1999) show how to fit an affine model via Kalman filtering. On this latter subject, Harvey (1989) and Commandeur and Koopman (2007) are key references. All in all, Fabozzi (2002) effectively examines the term structure together with related interest rate topics.

From a robust statistics perspective, Kharin (2013) focuses on time series analysis and forecast. Some interesting forward search applications can be found in Riani (2004, 2009).

APPENDIX A. KALMAN FILTER FOR AFFINE TERM STRUCTURE MODELS

The state space form is applied to the interest rate time series $\mathbf{R}_t = (R_{1,t}, + \cdots + R_{NN,t})$. Where at each time $t = 1, \ldots, T$ we consider NN observations (i.e., NN is the number of nodes of the interest rate curve). The standard Kalman filter provides a recursive algorithm for computing the minimum mean squared error estimator of $\boldsymbol{\alpha}_t$ conditional on past realizations (i.e., past interest rates) $\mathbf{R}_1, \ldots, \mathbf{R}_{t-1}$. The mean squared error of $\mathbf{a}_{t|t-1}$ is as follows:

$$MSE(\mathbf{a}_{t|t-1}) = \mathbb{E}[(\mathbf{a}_{t|t-1} - \boldsymbol{\alpha}_t)(\mathbf{a}_{t|t-1} - \boldsymbol{\alpha}_t)' | \mathbf{R}_1, \ldots, \mathbf{R}_{t-1}] = \mathbf{P}_{t|t-1}.$$

The Kalman filter is the set of recursions summarized as follows:

$$
\begin{aligned}
\boldsymbol{v}_t &= \mathbf{R}_t - \mathbf{c}_t - \mathbf{Z}_t \mathbf{a}_{t|t-1}, & \mathbf{F}_t &= \mathbf{Z}_t \mathbf{P}_{t|t-1} \mathbf{Z}_t' + \mathbf{G}_t \mathbf{G}_t', \\
q_t &= q_{t-1} + \boldsymbol{v}_t' \mathbf{F}_t^{-1} \boldsymbol{v}_t, & \mathbf{K}_t &= (\mathbf{T}_t \mathbf{P}_{t|t-1} \mathbf{Z}_t' + \mathbf{H}_t \mathbf{G}_t') \mathbf{F}_t^{-1}, & (3.36) \\
\mathbf{a}_{t+1|t} &= \mathbf{T}_t \mathbf{a}_{t|t-1} + \mathbf{d}_t + \mathbf{K}_t \boldsymbol{v}_t, & \mathbf{P}_{t+1|t} &= \mathbf{T}_t \mathbf{P}_{t|t-1} \mathbf{T}_t' + \mathbf{H}_t \mathbf{H}_t' - \mathbf{K}_t \mathbf{F}_t \mathbf{K}_t',
\end{aligned}
$$

where, for the interpretation of all other elements, the reader may refer to Harvey (1989) and Commandeur and Koopman (2007).

Two key components are quite important for the entire process. On the one hand, the filter innovations (one-step-ahead prediction errors) are indicated by $\boldsymbol{v}_t$. On the other hand, their variance is $\mathbf{F}_t$. These two quantities form the necessary ingredients for the computation of the likelihood:

$$l(\hat{\boldsymbol{\Theta}}) = -0.5 NN \cdot T \cdot \ln(2\pi) - 0.5 \sum_{t=1}^{T} \left(\ln |\mathbf{F}_t| + \boldsymbol{v}_t' \mathbf{F}_t^{-1} \boldsymbol{v}_t \right), \quad (3.37)$$

where $\hat{\boldsymbol{\Theta}}$ is the vector containing all the parameters of the model, $\boldsymbol{v}_t$ is the one-step-ahead prediction error, and $|\mathbf{F}_t|$ is the determinant of $\mathbf{F}_t$.

The multivariate state space form for the Vasicek and CIR models with J factors has the following measurement equation:

$$
\underbrace{\begin{pmatrix} R_t(\tau_1) \\ \vdots \\ R_t(\tau_{NN}) \end{pmatrix}}_{\mathbf{R}_t} = \underbrace{\begin{pmatrix} \frac{B_1(\tau_1)}{\tau_1} & \cdots & \frac{B_J(\tau_1)}{\tau_1} \\ \vdots & \ddots & \vdots \\ \frac{B_1(\tau_{NN})}{\tau_{NN}} & \cdots & \frac{B_J(\tau_{NN})}{\tau_{NN}} \end{pmatrix}}_{\mathbf{Z}_t} \underbrace{\begin{pmatrix} \alpha_{t,1} \\ \vdots \\ \alpha_{t,J} \end{pmatrix}}_{\boldsymbol{\alpha}_t}
$$

$$
- \underbrace{\begin{pmatrix} \frac{A(\tau_1)}{\tau_1} \\ \vdots \\ \frac{A(\tau_{NN})}{\tau_{NN}} \end{pmatrix}}_{\mathbf{c}_t} + \underbrace{\begin{pmatrix} \eta_{t,1} \\ \vdots \\ \eta_{t,NN} \end{pmatrix}}_{\mathbf{G}_t \boldsymbol{\epsilon}_t}, \quad (3.38)
$$

where $\tau_{(.)}$ denotes the maturity node of the term structure (e.g., 3 months, 3 years, 5 years, 10 years), and the functional forms for $A(\tau)$ and $B(\tau)$ can be formulated, for the Vasicek model, as

$$A(\tau) = \Sigma_{j=1}^{J}\left(\frac{\gamma_j(B_j(\tau)-\tau)}{k_j^2} - \frac{\sigma_j^2 B_j^2(\tau)}{4k_j}\right), \tag{3.39}$$

$$B_j(\tau) = \frac{1}{k_j}[1 - e^{-k_j\tau}], \tag{3.40}$$

where

$$\gamma_j = k_j^2\left(\theta_j - \frac{\sigma_j v_j}{k_j}\right) - \frac{\sigma_j^2}{2}. \tag{3.41}$$

On the other hand, for the CIR model we have

$$A(\tau) = \Sigma_{j=1}^{J}\ln\left(\frac{2\gamma_j e^{\frac{(\gamma_j+k_j+v_j)\tau}{2}}}{(\gamma_j + k_j + v_j)(e^{\gamma_j\tau}-1)+2\gamma_j}\right)^{\frac{2k_j\theta_j}{\sigma_j^2}}, \tag{3.42}$$

$$B_j(\tau) = \frac{2(e^{\gamma_j\tau}-1)}{(\gamma_j + k_j + v_j)(e^{\gamma_j\tau}-1)+2\gamma_j}, \tag{3.43}$$

where

$$\gamma_j = \sqrt{(k_j + v_j)^2 + 2\sigma_j^2}. \tag{3.44}$$

For both models, $\sigma_j \geq 0$ is the volatility parameter of the process for the jth factor. The meaning of the other parameters, k_j, θ_j, and v_j, referring to each of the J factors, was described after Eq. (3.14). A diagonal covariance structure for the errors in the measurement equation $\mathbf{G}_t\mathbf{G}_t'$ is assumed to ensure the identification of model parameters and reduce the complexity of Broyden-Fletcher-Goldfarb-Shanno optimization.

The transition equation is as follows:

$$\underbrace{\begin{pmatrix} \alpha_{t+1,1} \\ \vdots \\ \alpha_{t+1,J} \end{pmatrix}}_{\alpha_{t+1}} = \underbrace{\begin{pmatrix} e^{-k_1\Delta t} & 0 & \cdots & 0 \\ 0 & e^{-k_2\Delta t} & \cdots & 0 \\ \vdots & \vdots & \ddots & \vdots \\ 0 & \cdots & \cdots & e^{-k_J\Delta t} \end{pmatrix}}_{\mathbf{T}_t} \underbrace{\begin{pmatrix} \alpha_{t,1} \\ \vdots \\ \alpha_{t,J} \end{pmatrix}}_{\alpha_t}$$

$$+ \underbrace{\begin{pmatrix} \theta_1(1 - e^{-k_1\Delta t}) \\ \vdots \\ \theta_J(1 - e^{-k_J\Delta t}) \end{pmatrix}}_{\mathbf{d}_t} + \underbrace{\begin{pmatrix} \psi_{t,1} \\ \vdots \\ \psi_{t,J} \end{pmatrix}}_{\mathbf{H}_t\epsilon_t}, \tag{3.45}$$

where Δ_t, for daily observations, is conventionally assumed to be 1/250.

According to the Vasicek model specification, $\mathbf{H}_t\mathbf{H}_t'$ is assumed to be diagonal with jth diagonal element

$$\mathbf{H}_t\mathbf{H}_t'(j) = \frac{\sigma_j^2}{2k_j}(1 - e^{-2k_j\Delta t}). \tag{3.46}$$

In the case of the CIR model this matrix depends on a state space process with jth diagonal element

$$\mathbf{H}_t\mathbf{H}_t'(j) = \frac{\theta_j\sigma_j^2}{2k_j}\left(1 - e^{-k_j\Delta t}\right)^2 + \frac{\sigma_j^2}{k_j}\left(e^{-k_j\Delta t} - e^{-2k_j\Delta t}\right)\alpha_{t-1,j}. \tag{3.47}$$

APPENDIX B. ROBUST KALMAN FILTER: A FORWARD SEARCH APPROACH TO ESTIMATE AFFINE TERM STRUCTURE MODELS

As described in Appendix in Chapter 2, outliers may affect the data on which the analysis is to be performed. Robust statistics supplies useful tools to identify and deal with atypical units. On this subject, it is worth exploring additional applications of the forward search (FS). The Kalman filter procedure is a useful framework to test its effectiveness.

Let us follow the same steps described in Chapter 2 for initialization, progression, and monitoring of the FS. The only element deserving additional care is the measure of unit closeness. The likelihood function outlined in Eq. (3.37) allows us to estimate the parameter vector $\hat{\Theta}_*^{(m)}$ at each step m of the search. The FS progresses by use of the squared Mahalanobis distance of each unit t computed as follows:

$$d_t(m*) = v_t(m*)'[s^2(m*)]^{-1}v_t(m*), \tag{3.48}$$

where $s^2(m^*)$ is estimated as

$$s^2(m*) = \frac{\sum_{t \in S_*^{(m)}} v_t(m*)v_t(m*)'}{m - 1}, \tag{3.49}$$

and $v_t(\cdot)$ is obtained by use of the so-called Kalman gain only for the units belonging to the subset (Harvey, 1989).

As per the FS analysis in Chapter 2, the focus is on the minimum distance of units outside the subset:

$$d*_{tmin}(m*) = \min[d_t(m*)] \quad t \notin S_*^{(m)}. \tag{3.50}$$

As in standard regression (Atkinson and Riani, 2000), this distance increases progressively during the search and its slope is higher in the last few steps. The monitoring of $d_{tmin}^*(m^*)$ helps us to have an idea about potential outliers.

Let us apply the state space representation described in Section 3.2.3 and detailed in Appendix A to generate four paths of 60 units each of the Vasicek one-factor model with parameter vector Θ_{sim}=(0.03,0.6,0.01,−0.002, 0.0081, 0.0090, 0.0012, 0.0091)'. More specifically, these parameters correspond to the long-run average interest rate θ, the speed adjustment κ, the volatility σ, the risk premium ν, and the diagonal elements of the error matrix. The resulting time series represent daily interest rates at maturities: 3, 5, 7, and 10 years. Model parameters are estimated for all maturities τ by use of Eq. (3.37). A first analysis conducted on uncontaminated data shows that the forward search effectively does not highlight any atypical observation. The next step is to verify its capability to detect outliers. On this subject, units 25 to 27 are contaminated on maturities of 3 and 5 years. The top-left panel in Fig. 3.6 shows three outlying trajectories $d_t(m^*)$. The top-right panel compares $d^*_{tmin}(m^*)$ against its corresponding confidence bounds (dotted lines) obtained by use of the analytical approximation proposed by Riani et al. (2009). Three units cross the upper envelope and another is on the border. The bottom-left and bottom-right panels compare $d^*_{tmin}(m^*)$ against the confidence envelopes at subset size $T - 3$ and

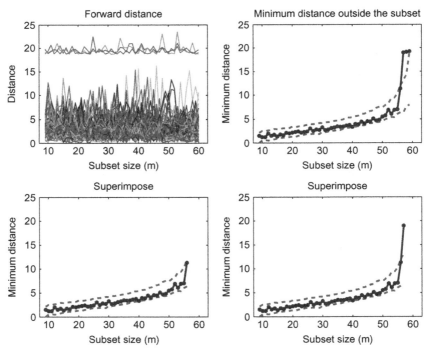

FIG. 3.6 Forward search analysis conducted on a Vasicek one-factor model with three contaminated units. *Top left*: Three outlying trajectories $d_t(m^*)$. *Top-right*: Comparison of $d^*_{tmin}(m^*)$ against its corresponding confidence bounds *(dotted lines)*. *Bottom left*: Comparison of $d^*_{tmin}(m^*)$ against the confidence envelopes at subset size $T - 3$. *Bottom right*: Comparison of $d^*_{tmin}(m^*)$ against the confidence envelopes at subset size $T - 2$.

TABLE 3.15 Asset and Liability Structure by Pivotal Interest Rate ($ Billions)

	Assets					
	[0–1M]	(1M–3M]	(3M–6M]	(6M–12M]	(>12M)	Total
Pivotal interest rate						
3 months	2.00	4.00		5.00	4.00	15.00
6 months	3.00	5.00	20.00	5.00	10.00	43.00
12 months			10.00	5.00	10.00	25.00
3 years		7.00		4.00		11.00
5 years		4.00				4.00
7 years		2.00				2.00
≥10 years						
Total assets	5.00	22.00	30.00	19.00	24.00	100.00
	Liabilities					
	[0–1M]	(1M–3M]	(3M–6M]	(6M–12M]	(>12M)	Total
Pivotal interest rate						
3 months	20.00	10.00	10.00			40.00
6 months	15.00		10.00			25.00
12 months		15.00		20.00		35.00
3 years						
5 years						
7 years						
≥10 years						
Total liabilities	35.00	25.00	20.00	20.00		100.00

M, Month(s).

$T - 2$ respectively. The upper threshold is crossed at subset size $T - 2$, which shows that the forward search effectively identifies all three contaminated units. Bellini and Riani (2012) showed that the procedure can be effectively applied to real time series.

EXERCISES

Exercise 3.1 Let us consider the Italian banking system time series introduced in Example 3.4 (file Chap3rates.xlsx). Use MATLAB to estimate the model parameters of Eq. (3.9) both on loan and on deposit time series by use of the LIBOR as an explanatory variable.

Exercise 3.2 Compute the term structure of interest rates for each of the simulated scenario introduced in Example 2.4 in Chapter 2 by use of MATLAB. The term structure needs to be constrained both on the short-term interest rate node and on the long-term interest rate node according to Bank Alpha's stress testing scenario described in Chapter 2. When the simulated long-term interest rate is zero, introduce a floor on the CIR model such that the long-term interest rate cannot fall below half the current θ parameter of the model.

Exercise 3.3 Estimate the margin at risk in terms of ΔNII for the bank described in Table 3.15 by use of the following assumptions:

- Use the loan and deposit sensitivity estimates of Exercise 3.1 for assets and liabilities respectively.
- Apply a shock corresponding to the difference between the 0.75 and 0.25 quantiles of the term structure distribution fitted in Exercise 3.2.

Solutions are available at www.tizianobellini.com

REFERENCES

Acerbi, C., Tasche, D., 2002. On the coherence of expected shortfall. J. Bank. Finance 26 (7), 1487–1503.

Atkinson, A.C., Riani, M., 2000. Robust Diagnostic Regression Analysis. Springer, New York.

Bellini, T., Riani, M., 2010. Un modello statistico per l'analisi della dipendenza temporale dei tassi bancari dai tassi interbancari. Excel per la finanza e il management, Alpha Test, Milan, pp. 273–303.

Bellini, T., Riani, M., 2012. Robust analysis of default intensity. Comput. Stat. Data Anal. 56 (11), 3276–3285.

Bibby, B., Sørensen, M., 1995. Martingale estimation functions for discretely observed diffusion processes. Bernoulli 1, 17–39.

BIS, 2011. Basel III: A global regulatory framework for more resilient banks and banking systems. Bank for International Settlements, Basel, Switzerland.

BOE, 2015. Stress testing the UK banking system: key elements of the 2015 stress test. Bank of England Publications, London.

Bohn, A., Elkenbracht-Huizing, M., 2014. The Handbook of ALM in Banking. Risk Books, London.

Bolder, D., 2001. Affine term structure models: theory and implementation. Working Paper 2001–15. Bank of Canada.

Brigo, D., Mercurio, F., 2006. Interest Rate Models: Theory and Practice with Smile, Inflation and Credit. Springer, Berlin.

Choudhry, M., 2007. Bank Asset and Liability Management: Strategy, Trading, Analysis. Wiley Finance, Singapore.

Christoffersen, P., 2012. Elements of Financial Risk Management, 2nd ed. Academic Press, San Diego.

Commandeur, J.J.F., Koopman, S.J., 2007. An Introduction to State Space Time Series Analysis. Oxford University Press, Oxford.

Cox, J., Ingersoll, J., Ross, S., 1985. A theory of the term structure of interest rates. Econometrica 53, 385–407.

Diebold, F., Rudebusch, G., Aruoba, B., 2006. The macroeconomy and the yield curve: a dynamic latent factor approach. J. Econom. 131, 309–338.

Duan, J., Simonato, J., 1999. Estimating and testing exponential-affine term structure models by Kalman filter. Rev. Quant. Finance Account. 13, 111–135.

Duffie, D., Singleton, K., 2003. Credit Risk Pricing, Measurement and Management. Princeton University Press, Princeton.

EBA, 2016. 2016 EU-wide stress test. Methodological note.

Fabozzi, F., 2002. Interest Rate, Term Structure, and Valuation Modelling. Wiley, Hoboken.

FRB, 2015. Comprehensive capital analysis and review 2015: assessment framework and results. Board of Governors of the Federal Reserve System, Washington, DC.

Harvey, A.C., 1989. Forecasting, Structural Time Series Models and the Kalman Filter. Cambridge University Press, Cambridge.

Hull, J.C., 2012. Options, Futures and Other Derivatives, 8th ed. Pearson, Essex.

Kashyap, A.K., Rajan, R., Stein, J.C., 2002. Banks as liquidity providers: an explanation for the coexistence of lending and deposit-taking. J. Finance 57, 33–73.

Kharin, Y., 2013. Robustness in Statistical Forecasting. Springer, New York.

McNeil, A., Frey, R., Embrechts, P., 2015. Quantitative Risk Management: Concepts, Techniques and Tools. Princeton University Press, Princeton.

Riani, M., 2004. Extension of the forward search to time series. Stud. Nonlinear Dyn. Econom. 8, 1–23.

Riani, M., 2009. Robust transformations in univariate and multivariate time series. Econom. Rev. 28, 262–278.

Riani, M., Atkinson, A.C., Cerioli, A., 2009. Finding an unknown number of multivariate outliers. J. R. Stat. Soc. Ser. B 71, 201–221.

Vasicek, O., 1977. An equilibrium characterization of the term structure. J. Financ. Econ. 5, 177–188.

Yoshida, N., 1992. Estimation for diffusion processes from discrete observations. J. Multivar. Anal. 41, 220–242.

Chapter 4

Portfolio Credit Risk Modeling

Chapter Outline

The focus of this chapter is on the banking book. More specifically, the attention is on credits constituting the vast bulk of a commercial bank's assets. A brief introduction outlines the scope and the tools to assess expected losses, unexpected losses, and risk-weighted assets (RWAs).

A primary goal is to describe portfolio models. The aim is to represent a loss distribution over a given time horizon (e.g., 1 year). This framework is at the heart of the internal ratings-based (IRB) capital requirement.

A distinction is made between advanced and standardized methods to estimate RWAs. On this subject, a bank may use IRB techniques to feed the RWA formula. In contrast, a standardized approach applies a simplified weighting scheme to a bank's assets.

A statistical framework is proposed to connect macroeconomic variables and credit risk parameters. This relationship is functional to stress testing and is crucial to estimate loan impairment charges.

Stress Testing and Risk Integration in Banks. http://dx.doi.org/10.1016/B978-0-12-803590-0.00002-1

Finally, a model combining portfolio modeling and stress testing techniques is shown. It paves the way to the risk integration mechanism described in the final two chapters.

KEY ABBREVIATIONS AND SYMBOLS

$A_{i,s,t}$ customer i in sector s asset exposure at time t

$CLoss$ portfolio credit loss

$LGD_{i,s,t}$ loss given default for the only facility of customer i in sector s at time t

$LGD_{pt,t}$ loss given default for a given product type pt at time t

$LGD_{pt,\Delta}$ loss given default shocked according to macroeconomic dynamics

$PD_{i,s|x_{scen}}$ conditional probability of default

$PD_{i,s,\Delta}$ probability of default shocked according to macroeconomic dynamics

$\mathbf{x}_t$ vector of macroeconomic variables at time t

$\mathbf{x}_\Delta$ vector of shocked macroeconomic variables

4.1 INTRODUCTION

In this chapter the banking book is explored by our focusing on the credit portfolio. In more detail, the aim is to investigate the loss distribution for a portfolio of credits. Such a distribution is built on the principle that the customer i may be insolvent and, in the case of default, its exposure may not be fully recovered. In this regard, the probability of default (PD_i) represents the likelihood of the insolvency. The loss given default (LGD_i) accounts for the unrecovered portion of an asset exposure (A_i).[1]

Alternative portfolio approaches have been developed in the recent past (Koyluoglu and Hickman, 1998). A credit loss is usually derived from the product of a given default indicator $(\mathbb{1}_{i,def})$ times LGD_i and A_i. Some models rely on a closed formula (Gordy, 2003) but more commonly a Monte Carlo simulation is used to derive such a distribution. For each simulation $g = 1, \ldots, G$, the portfolio credit loss $(CLoss_g)$ is computed as follows:

$$CLoss_g = \sum_{i=1}^{n} \mathbb{1}_{i,def,g} A_i LGD_i, \qquad (4.1)$$

whereby one may repeat the random process and obtain a distribution with the shape shown in Fig. 4.1. This curve highlights a typical asymmetric profile where small losses have higher chance than greater losses. At the same time,

---

1. This notation, instead of the more commonly used EAD_i, is used for alignment with Chapter 3 and the following chapters.

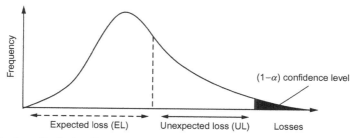

FIG. 4.1 Loss distribution. Expected loss versus unexpected loss.

Fig. 4.1 points out the distinction between expected loss and unexpected loss. The latter is obtained as a difference between a given quantile of the distribution (i.e., credit value at risk $VaR_{credit,(1-\alpha)}$) and the expected loss.

Section 4.2 focuses on the most typical portfolio credit models. We explore a common framework by aiming to highlight their affinities. CreditMetrics is investigated as representative of the so-called structural approach. In contrast, an intensity-based framework is scrutinized by use of copulas.

The following step is to investigate capital requirements and RWAs due to the credit risk. On this subject a distinction is made between standardized and IRB approaches. Section 4.3 shows that the unexpected loss concept depicted in Fig. 4.1 is vital for the IRB framework.

Section 4.4 describes the techniques most commonly applied to bridge the external economy and credit risk parameters, by aiming to specify a transmission process linking macroeconomic variables and internal risks.

Section 4.5 anticipates some of the key ideas driving the risk integration process described in the final two chapters.

From a toolkit perspective, a series of statistical methods are involved throughout the chapter. In particular, some basic probability theory concepts pave the way for the analysis of structural (probit) and logit probability of default models. Monte Carlo simulations together with copulas are used in almost all sections. Finally, a brief regulatory investigation relies on some introductory concepts of the Basel Accord framework.

4.2 CREDIT PORTFOLIO MODELING

In the mid-1990s, CreditMetrics (CreditMetrics, 1997), CreditRisk$^+$ (CSFB, 1997), Credit Portfolio View (Wilson, 1997), and other portfolio models became practical risk management tools for the major banks across the world. Each of these models pursues the goal of deriving a loss distribution on which to compute a synthetic measure of risk (e.g., $VaR_{credit,(1-\alpha)}$). According to Koyluoglu and Hickman (1998), the differences among these approaches can be reconciled in a more general framework. The regulatory IRB formula (Gordy, 2003) can also be examined through this paradigm.

4.2.1 Credit Loss Distribution

In line with the asset and liability framework described in Chapter 3, let us split the N assets of a bank $\mathbf{A}_t = (A_{1,t}, \ldots, A_{N,t})'$ by distinguishing between credits and noncredits. Here $n \leq N$ of these assets are credits (i.e., $\mathbf{A}_{credit,t} = (A_{1,t}, \ldots, A_{n,t})'$). Additionally, the assumption that each customer has only one facility (product) holds. Therefore the subscript i represents both the customer and the financial instrument. This hypothesis was introduced in Chapter 3 and applies throughout the book.

Portfolio models usually assume LGD_i and A_i as given. Nonetheless, randomness should be introduced without the overall framework being changed. In contrast, the probability of default is stochastic and is modeled by use of the following three components: unconditional default probability, creditworthiness index, and a framework linking default probabilities to the creditworthiness index.

These are the key ingredients of a framework representing the common elements of all major portfolio models. In what follows, a more detailed discussion is provided by our highlighting that each customer i is assumed to run a business belonging to a given sector s.

- **Unconditional default probability.** The unconditional 1-year (1_y) default probability for a bank's customer is defined as follows:

$$PD_{i,s,t} = P(\tau_{i,s} \leq 1_y), \tag{4.2}$$

where $\tau_{i,s}$ denotes the default time occurrence.

The 1-year holding period assumption is rooted in the conventional time needed to modify a credit portfolio composition to offset risky positions (e.g., via securitization or loan credit policy).

- **Creditworthiness index.** The second component is a creditworthiness index $\Psi_{i,s,t}$ through which the probability of default of customer i (operating in sector s) is linked to external variables. More precisely, the relationship between the creditworthiness index and a vector of macroeconomic variables $\mathbf{x}_t = (x_{1,t}, \ldots, x_{p,t})'$ as follows:

$$\Psi_{i,s,t} = \beta_{0,s} + \beta_{1,i,s}x_{1,t} + \cdots + \beta_{p,i,s}x_{p,t} + \sigma_{i,s}\epsilon_{i,s,t}, \tag{4.3}$$

where $\beta_{1,i,s}, \ldots, \beta_{p,i,s}$ represent the sensitivity over the vector of macroeconomic variables $\mathbf{x}_t = (x_{1,t}, \ldots, x_{p,t})'$.

All debtors belonging to a sector (s) are commonly assumed to have the same sensitivity over the above set of macroeconomic variables. Additionally, they share the same sector volatility (σ_s). Each debtor's idiosyncratic component is represented through the error term $\epsilon_{i,s,t} \sim N(0,1)$. The following assumption is also considered to ensure the random variable $\Psi_{i,s,t}$ has unit variance:

$$\sigma_{i,s} = \sqrt{1 - \beta_{1,i,s}^2 + \cdots + \beta_{p,i,s}^2}. \tag{4.4}$$

Example 4.1 outlines how to fit the creditworthiness index $\Psi_{s,t}$ represented by a sector default rate. The quarterly UK macroeconomic time series analyzed in Chapter 2 is used as explanatory variable. For the sake of simplicity, the analysis is performed on nonstandardized time series.

Example 4.1 $\Psi_{s,t}$ Fitting

The default rate time series is used as a creditworthiness index $\Psi_{s,t}$ for the UK corporate sector during the period from 2000 to 2013. The linear model outlined in Eq. (4.3) relies on the UK quarterly macroeconomic variables summarized in Table 4.1.

TABLE 4.1 List of UK Macroeconomic Variables Used to fit the Creditworthiness Index $\Psi_{s,t}$

Descriptions	Symbols and Analytical Formulas
Real output	$y_t = \ln(GDP_t/CPI_t)$
Inflation	$\Delta CPI_t = \ln(CPI_t/CPI_{t-1})$
Real equity price	$eq_t = \ln(EQ_t/CPI_t)$
Real exchange rate	$er_t = \ln(ER_t/CPI_t)$
Short-term interest rate	$r_t^{ST} = 0.25\ln(1 + R_t^{ST}/100)$
Long-term interest rate	$r_t^{LT} = 0.25\ln(1 + R_t^{LT}/100)$

The following MATLAB code is used to perform the regression. A few of the variables in Table 4.1 are included in the final model. Very good fitting is ensured by the inclusion of y_t, ΔCPI_t, and r_t^{ST}.

```
% File upload
mac=xlsread('Chap4PD.xlsx');
DR=mac(:,1);  GDP=mac(:,2); CPI=mac(:,3);
EQ=mac(:,4);  ER= mac(:,5); RS =mac(:,6);
RL=mac(:,7);
% 1. Macroeconomic variables included in the model
XX=[GDP(1:end-4,:) CPI(5:end,:) RS(5:end,:)];
% 2. Fitting
PD=fitlm(XX(1:end,:),DR(5:end,:));
FitPD=transpose(PD.Coefficients.Estimate)*...
transpose([ones(size(XX,1),1) XX]);
% Output
% Linear regression model:
%      y ~ 1 + x1 + x2 + x3
%
% Estimated Coefficients:
%              Estimate      SE       tStat      pValue
%
%             _____   _____   _____   _____
% (Intercept)  0.049166  0.0043062   11.417    6.9688e-15
% x1          -0.94798   0.25597     -3.7035   0.00057881
% x2           1.2506    0.4215       2.967    0.0048016
% x3          -3.2992    0.38101     -8.6592   3.8626e-11
```

(Continued)

Example 4.1 $\Psi_{s,t}$ Fitting—cont'd

```
% Number of observations: 49, Error degrees of freedom: 45
% Root Mean Squared Error: 0.0101
% R-squared: 0.837,  Adjusted R-Squared 0.826
% F-statistic vs. constant model: 77.2, p-value = 9.03e-18
```

The left-hand side of Fig. 4.2 outlines the dynamics of the macroeconomic time series included in the model together with the creditworthiness index. It is evident from this graph that a GDP reduction anticipates the peak of $\Psi_{s,t}$ (after the 2007–09 crisis). For this reason a lagged GDP growth is used as the explanatory variable. The right-hand side of Fig. 4.2 compares actual and fitted time series ($R^2 = 0.83$).

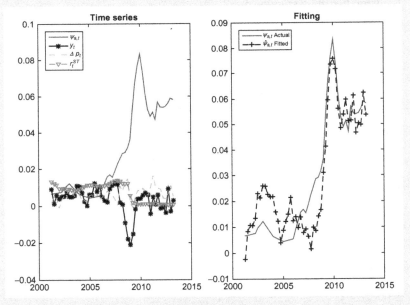

FIG. 4.2 Time series included in the model and $\Psi_{s,t}$ (*left*). Comparison of actual $\Psi_{s,t}$ and fitted $\hat{\Psi}_{s,t}$ (*right*).

- **Framework to link default probabilities to a creditworthiness index.** In what follows, two alternative ways to link default probabilities to a creditworthiness index are scrutinized. A distinction is made between a structural model rooted in the idea developed by Merton (1974) and the intensity approach.[2]

---

2. Here, intensity is used in a broad sense. For more details, see Section 4.2.3.

− **Structural model.** According to a structural model, default occurs when the asset value of a firm falls below the value of its liabilities. Asset standard returns represent the creditworthiness index $\Psi_{i,s,t}$ of Eq. (4.3). Hence default occurs when $\Psi_{i,s,t}$ falls below a given threshold as shown below:

$$P(\Psi_{i,s,t} < \alpha_{i,s,t}) = \Phi(\alpha_{i,s,t}) = PD_{i,s,t}, \qquad (4.5)$$

where Φ is the standard normal cumulative distribution function (CDF). One generally calculates the unconditional threshold by taking the inverse of Eq. (4.5). In other words, $\alpha_{i,s,t} = \Phi^{-1}(PD_{i,s,t})$, where Φ^{-1} stands for the inverse normal CDF. Consequently, the following indicator function summarizes the structural (probit) model framework:

$$\mathbb{1}_{i,s,t,def}^{Probit} = \begin{cases} 1 & \text{for} \quad \Psi_{i,s,t} \leq \Phi^{-1}(PD_{i,s,t}), \\ 0 & \text{for} \quad \Psi_{i,s,t} > \Phi^{-1}(PD_{i,s,t}). \end{cases} \qquad (4.6)$$

This indicator function and the corresponding conditional probability of default are a key element of the credit loss distribution. The latter is the paradigm on which the IRB regulatory formula relies (see Section 4.3) and is at the heart of the credit stress testing framework outlined in Section 4.4. More specifically, the conditional probability of default (on a given scenario, $\mathbf{x}_{scen}$) can be written as follows:

$$\begin{aligned} PD_{i,s,t|\mathbf{x}_{scen}}^{Probit} &= P\left(\Psi_{i,s,t} \leq \alpha_{i,s,t}|\mathbf{x}_{scen}\right) \\ &= P\left(\beta_{0,s} + \beta_{1,s}x_{1,t} + \cdots + \beta_{p,s}x_{p,t} + \sigma_s\epsilon_{i,s,t} \leq \alpha_{i,s,t}|\mathbf{x}_{scen}\right) \\ &= P\left(\epsilon_{i,s,t} \leq \frac{\alpha_{i,s,t} - (\beta_{0,s} + \beta_{1,s}x_{1,t} + \cdots + \beta_{p,s}x_{p,t})}{\sigma_s}|\mathbf{x}_{scen}\right) \\ &= \Phi\left(\frac{\alpha_{i,s,t} - (\beta_{0,s} + \beta_{1,s}x_{1,t} + \cdots + \beta_{p,s}x_{p,t})}{\sigma_s}|\mathbf{x}_{scen}\right) \\ &= \Phi\left(\frac{\alpha_{i,s,t} - (\beta_{0,s} + \beta_{1,s}x_{1,scen} + \cdots + \beta_{p,s}x_{p,scen})}{\sigma_s}\right). \end{aligned} \qquad (4.7)$$

It is worth noting that the equation $P\left(\epsilon_{i,s,t} \leq \frac{\alpha_{i,s,t}-(\beta_{0,s}+\beta_{1,s}x_{1,t}+\cdots+\beta_{p,s}x_{p,t})}{\sigma_s}\right)$ $= \Phi\left(\frac{\alpha_{i,s,t}-(\beta_{0,s}+\beta_{1,s}x_{1,t}+\cdots+\beta_{p,s}x_{p,t})}{\sigma_s}\right)$ relies on the fact that a CDF stands for the probability that a random variable assumes values lower or equal to a given realization.

− **Intensity approach.** Many authors, among which Wilson (1997), represent the probability that a debtor i belonging to sector s will default within a certain period (e.g., 1 year) as follows:

$$PD_{i,s,t} = \frac{1}{1 + e^{-\Psi_{i,s,t}}}. \qquad (4.8)$$

Then, one may consider the following default indicator function:

$$\mathbb{1}_{i,s,t,def}^{Logit} = \begin{cases} 1 & \text{for} \quad \frac{1}{1+e^{-\Psi_{i,s,t}}} \geq u_{i,s,t}, \\ 0 & \text{for} \quad \frac{1}{1+e^{-\Psi_{i,s,t}}} < u_{i,s,t}, \end{cases} \qquad (4.9)$$

where $u_{i,s,t} \in [0, 1]$ is a uniform random variable. The latter represents a casual event which can be drawn through Monte Carlo simulations. The following equation summarizes the logit conditional default probability

$$PD_{i,s|x_{scen}}^{Logit} = \frac{1}{1 + e^{-(\beta_{0,s} + \beta_{1,s}x_{1,scen} + \dots + \beta_{p,s}x_{p,scen})}}. \tag{4.10}$$

Appendix A compares linear and logit regression by highlighting the key features of this latter statistical technique.

Armed with the above credit risk structure based on a constant exposure, a loss given default (LGD), and stochastic default probability, all ingredients are available to estimate the portfolio loss distribution ($CLoss$). In line with the notation used in Chapter 3, $A_{i,s,t}$ denotes the exposure of debtor i belonging to sector s at time t. The loss associated with the default of debtor i is $LGD_{i,s,t}$ times its exposure $A_{i,s,t}$. The most intuitive way to use both structural and intensity models is to derive the loss distribution by generating Monte Carlo simulations. However, a closed form formula may be used to obtain the $CLoss$ distribution. This is the case of the IRB formula detailed in Section 4.3.2.

When following a simulation perspective, the default indicator $\mathbb{1}_{i,s,def}$ is generated through Monte Carlo simulations. Then, for each realization $g = 1, \dots, G$, the portfolio credit loss is computed as follows:

$$CLoss_g = \sum_{s=1}^{S} \sum_{i=1}^{n_s} \mathbb{1}_{i,s,t,def,g} A_{i,s,t} LGD_{i,s,t}, \tag{4.11}$$

where the double sum is meant to represent all customers of each sector (n_s), across all sectors S (i.e., $\sum_{s=1}^{S} \sum_{i=1}^{n_s} i = n$).

Example 4.2 illustrates how Eq. (4.11) works in practice.

Example 4.2 Credit Loss Simulation $CLoss_g$

A portfolio with $n = 5$ customers accounts for a total exposure of $5 million at time t. Table 4.2 summarizes how to compute the credit loss $CLoss_g$. It relies on a given random generation of the default indicator function $\mathbb{1}_{i,def,g}$, times the exposure A_i, times LGD_i.

TABLE 4.2 Credit Loss $CLoss_g$ ($ Millions)

i	$\mathbb{1}_{i,def,g}$	A_i	LGD_i	$CLoss_{i,g}$
1	0	1.50	20.00%	0.00
2	1	1.30	30.00%	0.39
3	0	1.20	40.00%	0.00
5	1	1.00	10.00%	0.10
Total		5.00		0.49

According to the above, simulation g is such that customer 1 does not default; thus the corresponding $CLoss_{1,g} = 0$. Customer 2 defaults, causing $CLoss_{2,g} = 0.39$. The third debtor does not default, while the fourth does with a corresponding $CLoss_{5,g} = 0.10$. The overall loss is $CLoss_g = 0.49$.

One derives the portfolio loss distribution by ordering $CLoss_g$. The expected loss corresponds to the average $CLoss$, and $VaR_{credit,(1-\alpha)}$ is computed as the $(1 - \alpha)$ quantile of the distribution. In contrast, the unexpected loss is derived as the difference between the credit value at risk $(VaR_{credit,(1-\alpha)})$ and the expected loss.

The next section outlines the so-called CreditMetrics portfolio model. This framework is widely used by banks. It paves the way to other credit risk practical implementations.

4.2.2 CreditMetrics

Example 4.3 is a convenient starting point to describe the CreditMetrics model.

Example 4.3 Introduction to CreditMetrics for a Single Loan Portfolio

Let us consider a BBB-rated loan with a bullet payoff of $5 million maturing in 5 years. It pays an annual coupon of $0.25 million.

Two choices are required for the purposes of this example. The first is to use seven rating categories from AAA to CCC (plus default) with a corresponding transition matrix. The second is to compute the risk over a 1-year horizon. The main purpose of the analysis is to point out all potential values of the loan at the end of the year by consideration of the following outcomes:

- The debtor stays at BBB at the end of the year.
- The debtor migrates up to AAA, AA, or A or down to BB, B, or CCC.
- The debtor defaults.

Each outcome has a different likelihood or probability of occurring. For now, the assumption is that the probabilities are known. That is, for a loan starting as BBB, the probabilities that it will end up in one of the seven rating categories (AAA through CCC) or defaults at the end of the year are known. These probabilities are shown in Table 4.3.

TABLE 4.3 Probability of Credit Rating Migrations in 1 Year for a BBB Loan

Year-End Rating	Probability (%)
AAA	0.05
AA	0.30
A	6.00
BBB	86.00
BB	5.00
B	2.00
CCC	0.45
Default	0.20

The present value of a loan depends upon the discount rate applied to its cash flows. Table 4.4 summarizes the 1-year forward zero curves by credit rating category.

(Continued)

Example 4.3 Introduction to CreditMetrics for a Single Loan Portfolio— cont'd

TABLE 4.4 Forward Zero Curves by Credit Rating Category

Rating	Year 1	Year 2	Year 3	Year 4
AAA	2.10%	2.67%	3.23%	3.62%
AA	2.25%	2.82%	3.38%	3.77%
A	2.42%	3.02%	3.63%	4.02%
BBB	2.90%	3.47%	4.05%	4.43%
BB	4.55%	5.02%	5.78%	6.27%
B	5.15%	6.12%	7.13%	7.62%
CCC	12.00%	12.20%	12.50%	12.85%

The value of the loan at the end of the first year in the case of no migration from rating BBB is as follows:

$$PV = \frac{0.25}{(1+2.90\%)^1} + \frac{0.25}{(1+3.47\%)^2} + \frac{0.25}{(1+4.05\%)^3} + \frac{5.25}{(1+4.43\%)^4} = 5.11.$$

Table 4.5 shows the loan value for all rating classes. The last column outlines the loss computed as a difference between the BBB and year-end potential rating classes.

TABLE 4.5 Loan Value and Loss for Each Migration Class ($ Millions)

Rating	Value	Loss
AAA	5.26	−0.15
AA	5.23	−0.12
A	5.19	−0.08
BBB	5.11	—
BB	4.79	0.32
B	4.58	0.54
CCC	3.83	1.28
Default	3.00	2.11

Remark. CreditMetrics applies a broader definition of loss compared to Eq. (4.1) by including adverse present value changes.

In what follows, the key steps of the CreditMetrics loss distribution estimation process are detailed.

- **Creditworthiness index.** A multifactor index representing normalized asset returns is used as in Merton (1974).
- **Rating thresholds.** Instead of a binary alternative (i.e., default against nondefault), the value of each loan relies on its year-end rating. Bands apply according to the cumulative probability of default. For example, a rating class CCC band corresponds to the cumulated probability of default inferred from the transition matrix as $\sum_{rating=default}^{CCC} Probability_{rating}$. The standard normal CDF (i.e., Φ^{-1}) is used as follows:

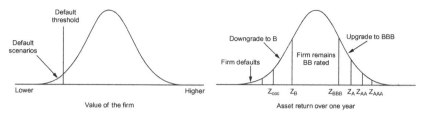

FIG. 4.3   Structural model default threshold compared with CreditMetrics rating bands.

$$Z_{default} = \Phi^{-1}(P_{default}),$$

$$Z_{CCC} = \Phi^{-1}\left(\sum_{rating=default}^{CCC} P_{rating} \right), \qquad (4.12)$$

$$Z_{B} = \Phi^{-1}\left(\sum_{rating=default}^{B} P_{rating} \right),$$

$$\dots,$$

where $Z_{(.)}$ represents the standard normal value as outlined on the right-hand side of Fig. 4.3.

- **Simulation.** The credit loss distribution is simulated following the next steps:
  - *Rating bands.* The normal inverse CDF is applied to assess each customer's year-end threshold band $Z_{(.)}$.
  - *Asset return simulation.* For a portfolio made up of n customers, g random returns are generated by use of a Cholesky decomposition of the customer's asset returns. This allows us to obtain an $n \times g$ matrix of correlated random generations.
  - *Year-end ratings.* The matrix of simulated asset returns matches each customer's threshold band $Z_{(.)}$ to obtain year-end ratings.
  - *Loss distribution.* The value of each loan is computed on the corresponding year-end rating class. Losses are obtained, for each simulation g, by comparison of the latter value against the no-rating-migration hypothesis.

Example 4.4 helps summarize the above process.

Example 4.4 CreditMetrics in Practice

Let us consider a portfolio with three loans. The first corresponds to the BBB rating credit examined in Example 4.3. Its value in the case of no rating migration is $5.11 million. Loan 2 has a value of $3.3 million if it does not move from its A rating class. Loan 3 has a value of $2.12 million because of its CCC rating. The portfolio has a value of $10.53 million when no migration occurs. However, fluctuations may happen, and Table 4.6 summarizes the value of loans in these events.

(Continued)

Example 4.4 CreditMetrics in Practice—cont'd

TABLE 4.6 Transition Probabilities (%) and Value ($ Millions)

Rating	Loan 1		Loan 2		Loan 3	
	Probability	Value	Probability	Value	Probability	Value
AAA	0.05	5.26	0.10	3.34	0.05	2.85
AA	0.30	5.23	2.00	3.33	0.45	2.83
A	6.00	5.19	90.00	3.30	1.00	2.81
BBB	86.00	5.11	5.00	3.25	1.50	2.77
BB	5.00	4.79	2.00	3.05	2.00	2.62
B	2.00	4.58	0.60	2.92	10.00	2.51
CCC	0.45	3.83	0.20	2.45	65.00	2.12
Default	0.20	3.00	0.10	1.80	20.00	1.20

The first step to compute rating thresholds is to estimate each customer's CDF as detailed in Table 4.7. Let us consider, for example, the default rating class for loan 1. The corresponding cumulative probability is 0.20%. The normal inverse CDF value is −2.88 (i.e., $\Phi^{-1}(0.002)$). For the rating class CCC, the cumulative probability is 0.20% + 0.45% = 0.65%, to which a normal inverse CDF value of −2.48 corresponds. This process is repeated for all rating classes as listed in Table 4.7.

TABLE 4.7 Rating Migration Thresholds Computed With Use of Eq. (4.12)

Rating	Loan 1		Loan 2		Loan 3	
	Cumulative Probability (%)	Threshold	Cumulative Probability (%)	Threshold	Cumulative Probability (%)	Threshold
AAA	100.00		100.00		100.00	
AA	99.95	3.29	99.90	3.09	99.95	3.29
A	99.65	2.70	97.90	2.03	99.50	2.58
BBB	93.65	1.53	7.90	−1.41	98.50	2.17
BB	7.65	−1.43	2.90	−1.90	97.00	1.88
B	2.65	−1.93	0.90	−2.37	95.00	1.64
CCC	0.65	−2.48	0.30	−2.75	85.00	1.04
Default	0.20	−2.88	0.10	−3.09	20.00	−0.84

A simulation process is performed by use of a Cholesky decomposition of the correlation matrix shown in Table 4.8.

TABLE 4.8 Correlation Matrix for the Loan Portfolio Under Analysis

	Loan 1	Loan 2	Loan 3
Debtor 1	1.00	0.40	0.10
Debtor 2	0.40	1.00	0.30
Debtor 3	0.10	0.30	1.00

As a following step, asset returns are generated on the basis of a multivariate normal variable as detailed in the three left-hand columns in Table 4.9 (i.e., 10 scenarios are considered). The following columns map each random generation in terms of the year-end rating class (on the basis of the above-mentioned bands).

Example 4.4 CreditMetrics in Practice—cont'd

TABLE 4.9 Scenario Generation and New Rating Derivation

	Random Generation			New Rating		
Scenario	Loan 1	Loan 2	Loan 3	Loan 1	Loan 2	Loan 3
1	1.50	−0.80	0.10	BBB	A	CCC
2	−2.10	−2.00	0.20	B	BB	CCC
3	−1.00	0.20	2.10	BBB	A	BBB
4	0.60	−0.10	−1.50	BBB	A	Default
5	0.40	−0.60	0.30	BBB	A	CCC
6	−0.10	0.50	−0.90	BBB	A	Default
7	0.80	1.50	−0.60	BBB	A	CCC
8	1.20	−0.70	−1.80	BBB	A	Default
9	1.80	2.00	1.10	A	A	B
10	0.10	−0.50	0.30	BBB	A	CCC

Finally, the portfolio value estimation concludes the process as detailed in Table 4.10. CLoss derives from the comparison of the portfolio value based on the year-end rating classes compared with $10.53 million corresponding to the no-rating-migration hypothesis.

TABLE 4.10 Portfolio Valuation for Each Scenario ($ Millions)

	New Rating			Value			Total Portfolio	
Scenario	Loan 1	Loan 2	Loan 3	Loan 1	Loan 2	Loan 3	Portfolio Value	CLoss
1	BBB	A	CCC	5.11	3.30	2.12	10.53	—
2	B	BB	CCC	4.58	3.05	2.12	9.75	−0.78
3	BBB	A	BBB	5.11	3.30	2.77	11.18	0.65
4	BBB	A	Default	5.11	3.30	1.20	9.61	−0.92
5	BBB	A	CCC	5.11	3.30	2.12	10.53	—
6	BBB	A	Default	5.11	3.30	1.20	9.61	−0.92
7	BBB	A	CCC	5.11	3.30	2.12	10.53	—
8	BBB	A	Default	5.11	3.30	1.20	9.61	−0.92
9	A	A	B	5.19	3.30	2.51	11.00	0.46
10	BBB	A	CCC	5.11	3.30	2.12	10.53	—

Starting from the CLoss distribution, unexpected loss is computed as the difference between a given percentile (e.g., 99.9%) and the corresponding expected loss (as illustrated in Fig. 4.1).

Example 4.5 shows an application of the CreditMetrics R package to a small credit portfolio.

Example 4.5 R CreditMetrics

Let us consider the CreditMetrics R package to derive the loss distribution of a portfolio made up of five loans with ratings BB, B, CCC, A, and B and corresponding exposure of $40 million, $100 million, $50 million, $300 million, and $150 million. The transition matrix is shown in Table 4.11.

(Continued)

Example 4.5 R CreditMetrics—cont'd

TABLE 4.11 One-Year Transition Matrix (%)

	AAA	AA	A	BBB	BB	B	CCC	D
AAA	90.81	8.33	0.68	0.06	0.08	0.02	0.01	0.01
AA	0.70	90.65	7.79	0.64	0.06	0.13	0.02	0.01
A	0.09	2.27	91.05	5.52	0.74	0.26	0.01	0.06
BBB	0.02	0.33	5.95	85.93	5.30	1.17	1.12	0.18
BB	0.03	0.14	0.67	7.73	80.53	8.84	1.00	1.06
B	0.01	0.11	0.24	0.43	6.48	83.46	4.07	5.20
CCC	0.21	0.00	0.22	1.30	2.38	11.24	64.86	19.79
D	0.00	0.00	0.00	0.00	0.00	0.00	0.00	100.00

The following R code performs the analysis based on the package CreditMetrics. The transition matrix and exposures feed the process together with a constant LGD of 40%, and interest rate 3%.

```
library(CreditMetrics)
lgd <- 0.40
# one year empirical migration matrix
rc <-       c("AAA", "AA", "A", "BBB", "BB", "B", "CCC", "D")
M <- matrix(c(90.81, 8.33, 0.68, 0.06, 0.08, 0.02, 0.01, 0.01,
0.70, 90.65, 7.79, 0.64, 0.06, 0.13, 0.02, 0.01,
0.09, 2.27, 91.05, 5.52, 0.74, 0.26, 0.01, 0.06,
0.02, 0.33, 5.95, 85.93, 5.30, 1.17, 1.12, 0.18,
0.03, 0.14, 0.67, 7.73, 80.53, 8.84, 1.00, 1.06,
0.01, 0.11, 0.24, 0.43, 6.48, 83.46, 4.07, 5.20,
0.21, 0, 0.22, 1.30, 2.38, 11.24, 64.86, 19.79,
0, 0, 0, 0, 0, 0, 0, 100
)/100, 8, 8, dimnames = list(rc, rc), byrow = TRUE)
cr.spread<-cm.cs(M, lgd)
# function cm.ref
r <- 0.03
ead <- c(40, 100, 50, 300, 150)
rating <- c("BB", "B", "CCC", "A", "B")
ref.val<-cm.ref(M, lgd, ead, r, rating)
ref.val$constVal
ref.val$constPV
```

The simulation process uses 1000 normal random generations, based on a given correlation matrix. For each of them a gain is computed. *CLoss* corresponds to gain with opposite sign. Fig. 4.4 highlights that small gains (negative losses) occur in around 700 of the 1000 simulations. The maximum *CLoss* is around $120 million out of a total portfolio face value of $640 million.

```
# function cm.rnorm.cor
N <- 5
n <- 1000
```

Example 4.5 R CreditMetrics—cont'd

```
firmnames <-
c("firm BB", "firm B", "firm CCC", "firm A", "firm B")
# correlation matrix
rho <- matrix(c( 1.0, 0.4, 0.6, 0.2, 0.1,
0.4, 1.0, 0.5, 0.3, 0.2,
0.6, 0.5, 1.0, 0.8, 0.7,
0.2, 0.3, 0.8, 1.0, 0.6,
0.1, 0.2, 0.7, 0.6, 1.0),
5, 5, dimnames = list(firmnames, firmnames),
byrow = TRUE)
rand.cor<-cm.rnorm.cor(N, n, rho)
# function cm.state
state.space<-cm.state(M, lgd, ead, N, r)
n=1000
# function cm.quantile
cm.quantile(M)
# function cm.val
val<-cm.val(M, lgd, ead, N, n, r, rho, rating)
val
# function cm.gain
gain<-cm.gain(M, lgd, ead, N, n, r, rho, rating)
gain
# hist
#n=10000
gain<-cm.gain(M, lgd, ead, N, n, r, rho, rating)
hist.gain<-hist(-gain, col="steelblue4",
main="CLoss Distribution",
xlab="CLoss", ylab="frequency")
```

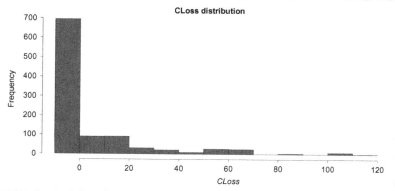

FIG. 4.4 *CLoss* distribution ($ millions).

In the next section a copula-based approach is introduced to obtain the credit loss distribution by use of a logit probability of default modeling.

4.2.3 Credit Portfolio Modeling With Copulas

The previous section highlighted how the CreditMetrics model enhanced the structural approach rooted in the idea of linking a firm's default to the relationship between its asset and liability values. In the recent past, some authors, for example, Schonbucher and Schubert (2001) and Duffie and Singleton (2003), proposed an alternative way to model portfolio losses. Their credit deterioration process relies on a function representing the intensity of default and another one merging debtors to account for their interdependencies. A probability of default framework captures the intensity, while a copula catches default linkages.

A copula function C is the joint distribution of a set of S uniform random variables $\mathbf{u} = (u_1, \ldots, u_S)$. A copula allows us to separate the marginal distribution from the function representing the dependence structure (Cherubini et al., 2004). The choice of the copula does not constrain the choice of the marginal distributions. Sklar (1959) showed that any multivariate distribution function F can be written in the form of a copula. According to Sklar's theorem, let G be an S-dimensional CDF with marginal distribution functions $F_1, \ldots, F_S$. Then there exists an n-dimensional copula C such that

$$G(y_1, \ldots, y_S) = C[F_1(y_1), \ldots, F_n(y_S)]. \tag{4.13}$$

If each function $F(\cdot)$ is continuous, then the copula C is unique. Skalar's theorem has an important corollary. Let G and C be, respectively, an S-dimensional distribution function with continuous univariate marginals $(F_1, \ldots, F_S)$ and an S-dimensional copula function. Then, for any $\mathbf{u} \in [0, 1]^S$, the following holds:

$$G(F_1^{-1}(u_1), \ldots, F_S^{-1}(u_S)) = C(u_1, \ldots, u_S), \tag{4.14}$$

where $F_s^{-1}(\cdot)$ denotes the inverse of the CDF.

Fig. 4.5 outlines copula random outcomes based on normal and Student T multivariate joint distributions. The first row highlights how normal realizations vary when the correlation parameter is increased: $0.1, 0.5$, and 0.9. A correlation increase causes the cloud to narrow along the diagonal. The second row highlights the higher frequency of off-diagonal observations due to the fat-tail Student T distribution.

In practice, the multivariate Student T copula is widely used for credit portfolio modeling. This is because of its fat-tail shape and because it is easy to treat even in the case of a large number of sectors S (in contrast, for example, to Archimedean copulas, for which very complicated procedures are required when $S > 2$). In the case of a Student T copula, let ρ be a symmetric, positive definite matrix with $\text{diag}(\rho) = (1, \ldots, 1)'$. Additionally, let $T_{\rho,\upsilon}$ be the standardized multivariate Student T distribution with correlation matrix ρ and υ degrees of freedom. The multivariate Student T copula function is as follows:

$$C(u_1, \ldots, u_S | \rho, \upsilon) = T_{\rho,\upsilon}(t_\upsilon^{-1}(u_1), \ldots, t_\upsilon^{-1}(u_S)), \tag{4.15}$$

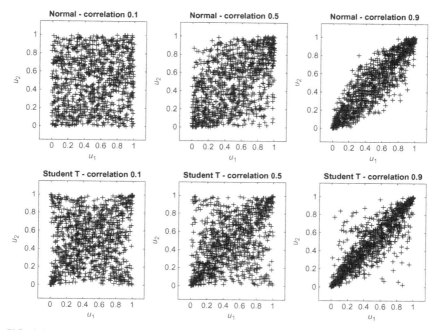

FIG. 4.5 Normal and Student T random copula simulation comparison for different correlation parameters (i.e., 0.1, 0.5, and 0.9).

where $t_\upsilon^{-1}(u_s)$ denotes the inverse of the Student T CDF for the sector s.

It is worth noting that outlying units may affect copula parameter estimates. Appendix B describes how to face this issue by use of the forward search procedure of Atkinson et al. (2004).

Starting from this representation, we can now easily outline how a portfolio model based on copulas works. A random number $u_{i,s,t}$ is generated for debtor i in sector s at time t by use of a given copula. Then a simple two-state model (i.e., default versus nondefault) is required as detailed in Eq. (4.9). Sector correlations are estimated on the basis of the creditworthiness index $\Psi_{s,t}$ time series. As anticipated in Section 4.2.1, an assumption is commonly made: each debtor i belonging to sector s is perfectly correlated to the other debtors in the same sector s. Then it is sufficient to generate a vector of uniform variables $\mathbf{u}_t = (u_{1,t}, \ldots, u_{S,t})'$ from a given copula. Eventually, the default indicator function $\mathbb{1}_{i,s,t,def}$ is obtained by comparison of $PD_{i,s,t}$ and $u_{s,t}$. Default occurs when $PD_{i,s,t} \geq u_{s,t}$. The vector $\mathbf{u}_t = (u_{1,t}, \ldots, u_{S,t})'$ is generated for each time t of a multiperiod exercise.

Example 4.6 highlights the role of copulas in assessing portfolio risk.

Example 4.6 The Role of Copulas in Portfolio Loss Computation

A $10 million small- to medium-sized Italian corporate portfolio accounts for 200 loans with exposure ranging from $40,000 and $60,000. Loans are spread among 20 different sectors. Default probabilities range from 0.3% to 15%, while LGDs are distributed within the interval from 10% and 30%. Let us estimate the loss distribution and compute $VaR_{credit,(1-\alpha)}$ and $ES_{credit,(1-\alpha)}$ by use of a normal copula with correlation parameter 0.10 (i.e., low correlation). The MATLAB code to derive the portfolio loss distribution follows.

```
ptfgran=dataset('XLSFile','copula20.xlsx');
ptfgran=ptfgran(:,{'SECTOR','ID','EXPOSURE','PD','LGD'});
% 2. Parameter set-up
%state1=1234567;
randn('state', 1234567);
ncustomer=size(ptfgran,1);
nsim=10000;
nsec=20;
corr=0.10;
for i=1:nsec
    for j=1:nsec
        if i==j
        corrgran(i,j)=1;
            else
        corrgran(i,j)=corr;
            end
        end
    end
end
% 3. Copula random simulation
rndgran = copularnd('Gaussian',corrgran, nsim)';
% 4. Database enrichement with copula simulations
ssgran=(1:1:nsec)';
rndgran1 = dataset({ssgran, 'SECTOR'},...
{rndgran(:,:),'NameRnd'});
joindbgran = join(ptfgran,rndgran1);
ptfcopulagran = double(joindbgran);
% 5. Default simulation and portfolio evaluation
simdefaultgran = NaN(ncustomer,nsim);
for i=1:ncustomer
    for j=1:nsim
    verif= ptfcopulagran(i,4)- ptfcopulagran(i,5+j);
        if(verif>=0)
        simdefaultgran(i,j)= (1-ptfcopulagran(i,5))*...
        ptfcopulagran(i,3); % 1-LGD
        else
        simdefaultgran(i,j)= 1*ptfcopulagran(i,3);
        end
```

Example 4.6 The Role of Copulas in Portfolio Loss Computation—cont'd

```
      end
end
ptfvaluegran =sum(simdefaultgran);
% 6. Loss computation
percen=0.999;
lossgran=sum(ptfcopulagran(:,3)) - ptfvaluegran;
ELgran=sum(ptfcopulagran(:,3).*ptfcopulagran(:,4)...
.*ptfcopulagran(:,5));
biasloss=(mean(lossgran)-ELgran)/ELgran;
qcopulagran=quantile(lossgran,percen);
ULcopulagran= qcopulagran - mean(lossgran);
ULcopulagran/sum(ptfcopulagran(:,3));
sortlossgran=sort(lossgran);
ES=mean(sortlossgran(1,percen*nsim:end));
```

The $163,290 simulated expected loss closely approximates the corresponding expected loss of $163,030 computed as a product of the exposure at default, LGD, and probability of default. Fig. 4.6 shows the loss distribution (losses less than $100,000 are excluded from the histogram). The $VaR_{99.9\%}$ of the portfolio is $835,380 and the $ES_{99.9\%}$ is $936,170. The unexpected losses based on the value at risk and expected shortfall are $672,090 $UL_{VaR_{99.9\%}}$ and $772,880 $UL_{ES_{99.9\%}}$ respectively . They account for 6.72% or 7.72% of the nominal portfolio value.

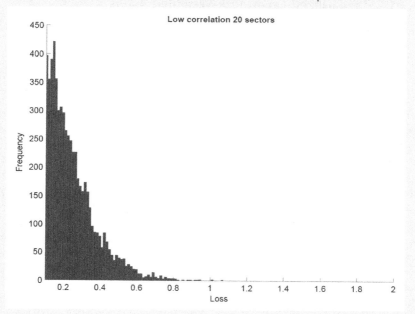

FIG. 4.6 Portfolio loss distribution computed by use of low correlation, $\rho = 0.10$, on a portfolio spread among 20 sectors ($ millions).

The next section investigates how to compute the regulatory RWAs. Two main approaches may be applied. On the one hand, the standardized method relies on a risk weight scheme based on regulatory exposure categories. On the other hand, the IRB method is rooted in a portfolio model aligned with the structural perspective discussed in Section 4.2.1.

4.3 CREDIT RISK-WEIGHTED ASSETS

As described in Chapter 1, the Basel II framework relies on three pillars: minimum capital requirements (pillar 1), regulatory supervision (pillar 2), and market disclosure (pillar 3). With respect to credit risk capital requirements, the Basel II Accord permits banks to adopt one of the following two methods: standardized approach or an IRB method. The latter provides two alternatives: foundation and advanced. Basel III Accord relies on the credit risk architecture defined under Basel II. Therefore, credit risk Basel II rules are still operative under Basel III.

In light of this distinction, the following sections summarize and explain the standardized approach and the IRB method. In particular, the latter will be analyzed through a portfolio modeling perspective as detailed in Section 4.2.

4.3.1 Standardized Credit Risk-Weighted Assets

According to the standardized approach, a given weight (W_{ss}) is applied to the sum of all instruments within the same risk category as follows:

$$RWA_{cr,std} = \sum_{ss=1}^{SS} \left(\sum_{i \in ss} EXP_{i,ss} \right) W_{ss}, \qquad (4.16)$$

where the subscript i stands for the customer, ss indicates a standardized credit risk category (a bank may have SS of these classes), and $EXP_{i,ss}$ denotes the exposure. A risk weight W_{ss} is applied to the sum of all exposures within ss.

The reason behind this approach is to couple the highest weights with the riskiest categories. No explicit portfolio modeling theory supports this scheme. However, weights represent a practical and prudential assessment of each group risk profile. It is worth noting that the Basel Committee on Banking Supervision (BIS, 2015) published a consultation paper to revise the standardized approach for credit risk. Its aim is to tackle some of the weaknesses in the current approach. Firstly, a reduction of the resilience on external credit ratings inspired the document. Then, additional granularity and risk sensitivity are required. Finally, the improvement of the comparability across different jurisdictions aims to avoid misalignment of treatment between standardized and IRB approaches.

In what follows, a summary of the standardized categories is outlined to highlight similarities among exposures within the same risk grade (BIS, 2006).

Additionally, product type and mitigation elements are considered to choose the weight W_{ss}.

- **Claims on sovereign states, public sector entities, and multilateral development banks.** A risk weight scheme is based on the rating assigned to that government by a recognized export credit agency (ECA). Starting from a minimum 0%, weights rise until 150% according to the rating. The risk weight is 100% for nonrated exposures (i.e., exposures without ECA rating).
- **Claims on banks and securities firms.** The Basel II Accord gives two options. According to the first option, claims on banks and securities firms are assigned the risk weight category below the country's risk weight category. The second option is to weight banks and securities firms on the basis of an external credit assessment scores.
- **Claims on corporations.** A risk weight is assigned according to the rating of the corporation or the asset. An external credit assessment institution (ECAI) that satisfies certain criteria described in the Basel II accord (BIS, 2006) is required to attribute such a rating. Starting from a minimum 20%, weights rise until 150% according to the rating. For nonrated exposures (i.e., without ECAI rating), the risk weight is 100%.
- **Claims included in the regulatory retail portfolios.** Loans to individuals and small businesses, including credit card loans and loans to small business entities, have a 75% weight. In this regard the definition of a small business is subject to regulatory discretion. All in all, a retail portfolio is characterized by no concentration (i.e., no single asset exceeds 0.2% of the entire retail portfolio, and no loan exceeds €1 million).
- **Claims secured by residential properties.** Residential real estate loans have a 35% weight.
- **Claims secured by commercial real estate.** In general, a 100% weight applies to loans secured by commercial real estate. However, the Basel II Accord permits regulators the discretion to assign mortgages on commercial properties a 50% weight subject to certain prudential limits.
- **Past due loans.** The unsecured portion of any loan that is past due for more than 90 days has a 150% weight when specific provisions are less than 20% of the outstanding loan. A 100% weight holds when specific provisions are greater than 20%.
- **High-risk categories.** A 150% weight is applied to claims on sovereign states, public sector entities, banks, and securities firms rated below B−.
- **Off–balance sheet items.** For the off-balance sheet exposures the Basel II Accord requires a conversion to credit exposure equivalents by use of credit conversion factors (e.g., securities used as collateral are converted to on-balance sheet assets by means of a 100% conversion credit conversion factor, self-liquidating trade letters collateralized by the goods being shipped are converted by means of a 20% factor, and so on).
- **Credit risk mitigation.** The standardized approach recognizes a wide range of risk mitigation techniques. In particular, for collateral, banks have two

options. Under the simple approach, a bank may adjust the risk weight for its exposure by using the appropriate weight for the supporting collateral instrument. The collateral must be marked to market and revalued at least every 6 months. A risk weight floor of 20% will also apply, unless the collateral is cash, or a government security, or belongs to some repo categories. Eligible collaterals include corporate debt instruments with high rating, equity securities traded on a main index, and government instruments. Under the second option, or comprehensive approach, the value of the exposure is reduced by a discounted value of the collateral. The Basel Accord provides the amount of the discount.

Example 4.7 summarizes how to compute RWAs by use of the standardized approach.

Example 4.7 Computation of RWAs by the Standardized Approach

Let us consider a $10 billion credit portfolio spread among the regulatory categories detailed in Table 4.12. For each group of exposures a specific risk weight is required to compute the corresponding RWAs. The last two columns summarize how to perform the calculation.

TABLE 4.12 Computation of Credit Risk-Weighted Assets ($ Billions) by the Standardized Approach

Portfolio	$\sum_{i \in ss} EXP_{i,ss}$	W_{ss} (%)	RWAs
Claims on corporations	4.00	100.00	4.00
Retail exposures	2.00	75.00	1.50
Residential real estate	1.00	35.00	0.35
Sovereign and central banks	3.00	20.00	0.60
Total	10.00		6.45

RWAs, Risk-weighted assets.

Basel capital requirement corresponds to 8% of the RWAs. Therefore 8% × $6.45 billion = $0.516 billion is the minimum capital to run the business referred to the credit portfolio under analysis. For a more detailed description, the reader may refer to Chapter 6.

The next section explores the IRB approach by starting from the portfolio modeling perspective introduced in Section 4.2.1.

4.3.2 Internal Ratings-Based Credit Risk-Weighted Assets

The regulatory IRB framework defines the minimum capital requirement for a bank by use of the unexpected loss described in Fig. 4.1. A Merton-like approach inspires the framework. More specifically, the so-called asymptotic single risk factor model infers the default distribution on the basis of a firm's asset value returns. On this subject, a parallel is evident with the creditworthiness index introduced in Section 4.2.1. The creditworthiness index ($\Psi_{i,s,t}$) corresponds to a normalized asset return $\xi_{i,s,t}$. In the IRB framework a single common factor ($\varkappa_t$) and an idiosyncratic noise component ($\epsilon_{i,s,t}$) drive this index:

$$\xi_{i,s,t} = \sqrt{\rho_{i,s}}\varkappa_t + \sqrt{1 - \rho_{i,s}}\epsilon_{i,s,t}, \tag{4.17}$$

where $\sqrt{\rho_{i,s}}$ is the correlation between asset returns and the common factor (Duffie and Singleton, 2003). $\varkappa_t$ and $\epsilon_{i,s,t}$ are i.i.d. $N(0, 1)$. Therefore $\xi_{i,s,t}$ has a standardized normal distribution.

A binary default random variable is the starting point of the model used for the IRB method as detailed below:

$$\mathbb{1}_{i,s,def}^{IRB} = \begin{cases} 1 & \text{for} \quad \xi_{i,s,t} \leq \Phi^{-1}(PD_{i,s,t}), \\ 0 & \text{for} \quad \xi_{i,s,t} > \Phi^{-1}(PD_{i,s,t}), \end{cases} \tag{4.18}$$

where Φ is the CDF of a standard normal distribution. The following equation summarizes the probability of default for a specific scenario $\varkappa_{scen}$:

$$
\begin{aligned}
P\left(\mathbb{1}_{i,s,def}^{IRB} = 1|\varkappa_{scen}\right) &= P\left(\xi_{i,s,t} \leq \Phi^{-1}(PD_{i,s,t})|\varkappa_{scen}\right) \\
&= P\left(\sqrt{\rho_{i,s}}\varkappa_t + \sqrt{1 - \rho_{i,s}}\epsilon_{i,s,t} \leq \Phi^{-1}(PD_{i,s,t})|\varkappa_{scen}\right) \\
&= P\left(\epsilon_{i,s,t} \leq \frac{\Phi^{-1}(PD_{i,s,t}) - \sqrt{\rho_{i,s}}\varkappa_t}{\sqrt{1 - \rho_{i,s}}}|\varkappa_{scen}\right) \\
&= \Phi\left(\frac{\Phi^{-1}(PD_{i,s,t}) - \sqrt{\rho_{i,s}}\varkappa_t}{\sqrt{1 - \rho_{i,s}}}|\varkappa_{scen}\right) \\
&= \Phi\left(\frac{\Phi^{-1}(PD_{i,s,t}) - \sqrt{\rho_{i,s}}\varkappa_{scen}}{\sqrt{1 - \rho_{i,s}}}\right).
\end{aligned}
\tag{4.19}
$$

Vasicek (2002) and Gordy (2003) extended the single asset model of Merton (1974) to a portfolio model by use of an infinitely granular portfolio where each debtor is independent of the others. The asymptotic single risk factor assumption informs the IRB formula used to assess capital requirements (BIS, 2006).

The IRB formula is derived as follows. Let us start from the exposure at default denoted as $A_{i,s,t}$ (where $\sum_{s=1}^{S} n_s = n$). Each of these assets represents a credit toward a given customer i belonging to a sector s. For each customer and facility the following credit risk parameters hold: $PD_{i,s,t}$, $LGD_{i,s,t}$, and $A_{i,s,t}$. Additionally, $\rho_{i,s}$ is the correlation between the asset i and the common factor $\varkappa$. The loss over a given time horizon h (i.e., 1 year) should be obtained through Monte Carlo simulations as detailed in Eq. (4.11). However, the assumption

$n \to \infty$ together with the infinite granularity of the portfolio allows us to obtain a closed-form expression to derive the $(1-\alpha)$ percentile of the loss distribution. More specifically, the $(1-\alpha)$ percentile of the conditional default probability described in Eq. (4.19) allows us to derive what follows:

$$
CLoss_{h,(1-\alpha)} = \sum_{s=1}^{S} \sum_{i=1}^{n_s} A_{i,s,t} LGD_{i,s,t} \Phi \left(\frac{\Phi^{-1}(PD_{i,s,t}) - \sqrt{\rho_{i,s}}\Phi(1-\alpha)}{\sqrt{1-\rho_{i,s}}} \right).
$$
(4.20)

The unexpected loss $(UL_{h,(1-\alpha)})$ is the difference between $CLoss_{h,(1-\alpha)}$ and the expected loss as detailed below:

$$
UL_{h,(1-\alpha)} = \sum_{s=1}^{S} \sum_{i=1}^{n_s} A_{i,s,t} LGD_{i,s,t} \Phi \left(\frac{\Phi^{-1}(PD_{i,s,t}) - \sqrt{\rho_{i,s}}\Phi(1-\alpha)}{\sqrt{1-\rho_{i,s}}} - PD_{i,s,t} \right).
$$
(4.21)

In Example 4.7 an 8% coefficient was used to compute the minimum capital requirement. Here the reciprocal of 8% (as well as an additional 1.06 overall adjustment) is applied to compute the RWAs by our starting from the above unexpected loss estimate:

$$
RWA_{cr,IRB} = 12.5 \cdot 1.06 \sum_{i=1}^{n} A_{i,s,t} LGD_{i,s,t} \Phi \left(\frac{\Phi^{-1}(PD_{i,s,t}) + \sqrt{\rho}\Phi^{-1}(0.999)}{\sqrt{1-\rho}} - PD_{i,s,t} \right) adj(M_{i,s,t}),
$$
(4.22)

where $A_{i,s,t}$ represents the exposure at default. This notation is used for alignment with the formalization used throughout the book to represent a customer's exposure. The multiplier $adj(M)$ penalizes exposures with high maturity (M).

The 0.999 percentile corresponds to $(1-\alpha)$ in Eq. (4.21), and the parameter ρ is defined according to the segment (i.e., large corporations, small- and medium-sized enterprises, etc.).[3]

It is worth noting that Eq. (4.22) is used for the nondefaulted portfolio only. Chapter 6 will describe the treatment for the defaulted portfolio. Additionally, a distinction holds between foundation and advanced IRB approaches. According to the foundation approach, a bank is authorized to use internally estimated default probabilities, while all other parameters are given (e.g., LGD of 45%). In contrast, an advanced IRB bank relies on internally estimated default probabilities, LGDs, and exposures at default.

Example 4.8 outlines how the IRB formula works in practice.

---

3. It is worth noting that in BIS (2006), the correlation parameter is indicated by R and different formulas apply according to the segment. For example, for corporate, sovereign, and bank exposures, $R = 0.12 \times (1 - e^{-50 \times PD})/(1 - e^{-50}) + 0.24 \times [1 - (1 - e^{-50 \times PD})/(1 - e^{-50})]$.

Example 4.8 IRB Approach for RWAs

Let us consider a portfolio with six customers each with $1 million exposure. Three of them belong to the small- and medium-sized enterprise segment (their turnover is €10 million), while the others are classified in the corporate portfolio (see Table 4.13). For both segments, debtors have default probabilities of 1%, 2%, and 3% and a fixed 40% LGD. The maturity is 5 years for all customers. A first comparison is between the RWAs computed for small- and medium-sized enterprise and corporate debtors. All other elements being equal, the value of the small- and medium-sized enterprise RWAs is lower because of the presumed low risk related to small companies. Additionally, it is useful to highlight that the weight resulting from the IRB regulatory formula is less than 100% only for customers 1 and 4 (i.e., default probability of 0.5%). In this regard, Fig. 4.7 shows that the IRB weighting scheme supports high-quality credits and penalizes low-quality ones.

TABLE 4.13 Bank Rho Credit Risk-Weighted Assets ($ Millions)

Customer	Segment	PD (%)	RWAs
1	SME	0.50	0.75
2	SME	2.00	1.10
3	SME	3.00	1.18
4	Corporate	0.50	0.93
5	Corporate	2.00	1.38
6	Corporate	3.00	1.50
Total			6.85

PD, Probability of default; RWAs, risk-weighted assets; SME, small- and medium-sized enterprise.

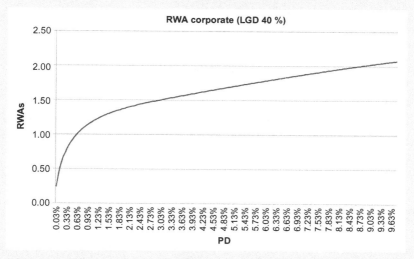

FIG. 4.7 Internal ratings-based risk-weighted assets *(RWAs)* curve for corporate exposure (for a loss given default of 40%). *PD*, Probability of default.

The next section focuses on Bank Alpha's RWAs for credit risk.

4.3.3 Bank Alpha's RWAs for Credit Risk

According to the balance sheet structure introduced in Chapter 3, Bank Alpha's overall loan portfolio is $70 billion net from provisions, while its value gross from provisions is $75 billion. A distinction is made between performing and nonperforming (defaulted) credits. The performing portfolio gross from provisions accounts for $62.40 billion, while the nonperforming portfolio accounts for $12.60 billion. Eq. (4.22) applies to the performing exposures (i.e., $62.40 billion).

Table 4.14 outlines that Bank Alpha's credit portfolio is split among corporate, retail, and other segments. In terms of exposure, the size of each subportfolio is similar. However, the corporate segment has the highest average default probability, and it has the lowest average LGD. A 43.74% retail portfolio LGD highlights the prevalence of unsecured operations compared with secured ones. The $54.80 billion RWA estimate relies on Eq. (4.22). This estimate will be used in Chapter 6 as a starting point to outline how to assess the RWAs under a stress testing scenario.

The next section describes how to link credit risk parameters to macroeconomic variables. It paves the way to the stress testing process that will be completed in Chapters 5 and 6. Furthermore, this method will be widely used for risk integration and reverse stress testing outlined in Chapters 7 and 8.

TABLE 4.14 Bank Alpha's Credit Risk Parameters, Risk-Weighted Assets, and Capital Requirements at t_0 ($ Billions)

	Average PD (%)	Average LGD (%)	Sum of Exposures	RWAs	Capital Requirements
Corporate	4.65	23.00	19.19	14.69	1.17
Retail	3.68	43.74	20.88	26.90	2.15
Others	2.14	25.34	22.33	13.21	1.06
Total	3.43	30.78	62.40	54.80	4.38

LGD, loss given default; PD, probability of default; RWAs, risk-weighted assets.

4.4 HOW TO LINK CREDIT RISK PARAMETERS AND MACROECONOMIC VARIABLES

Previous sections described how to assess portfolio credit risk. A few risk parameters were analyzed. It is now useful to detail how to link them to the external economy so as to perform a stress testing exercise. In this regard, two main approaches are considered. On the one hand, panel models allow the inclusion of macroeconomic variables directly into their estimates. On the other hand, a two-step process is used. Firstly, credit risk parameters are estimated for IRB purposes by focusing on a 1-year holding period (without explicitly accounting for macroeconomic variables). Then some adjustment is applied to account for external macroeconomic fluctuations.

In practice, banks do not usually implement new PD models for stress test only. The vast majority of banks develop models for IRB purposes and use them also for stress testing. Therefore the above two-step process applies where an additional modeling layer is required to link risk parameters to macroeconomic variables. The same applies to LGDs. Chapter 5 will embrace balance sheet projections and the corresponding credit exposure evolution.

4.4.1 Default Probability and Macroeconomic Variables

In line with the previously described two-step modeling approach, the first step to link default probabilities and macroeconomic variables is to highlight how the latter are estimated for IRB purposes. In this regard, banks have developed different approaches. In particular, expert assessments and statistical models are commonly combined to infer the likelihood of a customer becoming insolvent. Market data such as equity time series and credit risk spreads are used in conjunction with nonmarket data (e.g., balance sheet indices, credit bureau information, and the bank's internal data).

From a statistical point of view, a series of techniques have been scrutinized and adopted to assess default probabilities. A pioneer study by Altman (1968) showed how to use discriminant analysis in credit risk assessment. Neural networks, classification and regression trees, data envelopment analysis, and many other approaches appeared afterward. However, they all were characterized by difficulties in terms of implementation and opacity when interpretation was necessary. A simpler approach relies on the generalized linear model regression. This latter is the most commonly used method in risk management practice. This is because of both its easy implementation and its immediate interpretation. Appendix A shows the key characteristics of this approach by focusing on the logit link function. A comparison between logit and linear regression shows the key advantages of the generalized linear model. In addition, Exercise 4.1 enters into the details of the default probability estimation for a small sample of firms.

Originating from the logit model, the probability of default for debtor i operating in sector s conditioned on a specific realization of a creditworthiness index $\Psi_{s,t} \in [0, 1]$ is as follows:

$$PD_{i,s,t} = P\left(\mathbb{1}_{i,s,def} = 1|\Psi_{s,t}\right) = \frac{1}{1 + e^{-(\zeta_{i,s,t}|\Psi_{s,t})}}. \tag{4.23}$$

Here $\zeta_{i,s,t}$ depends on both microeconomic factors χ_l and the creditworthiness index $\Psi_{s,t}$ as described below:

$$\zeta_{i,s,t} = \eta_{i,s,0} + \eta_{i,s,1}\chi_{1,t} + \cdots + \eta_{i,s,k-1}\chi_{k-1,t} + \eta_{i,s,k}\Psi_{s,t} + \epsilon_{i,s,t}, \tag{4.24}$$

where $\epsilon_{i,s,t}$ is the error component.

It is worth mentioning that Eq. (4.23) highlights a more general frame compared with Eq. (4.8). Indeed, Eq. (4.8) directly links a debtor's default probability to a creditworthiness index, while in practice IRB default probabilities are estimated by use of Eq. (4.23).

The sector s creditworthiness index is modeled against macroeconomic variables as follows:

$$\Psi_{s,t} = \beta_{s,0} + \beta_{s,1} x_{1,t} + \cdots + \beta_{s,p} x_{p,t} + \epsilon_{s,t}, \tag{4.25}$$

where the vector $\mathbf{x} = (x_1, \ldots, x_p)'$ represents macroeconomic variables, and $\boldsymbol{\beta} = (\beta_{s,0}, \ldots, \beta_{s,p})'$ is the vector of coefficients and $\epsilon_{s,t}$ is the error term.

The stressed creditworthiness index is then projected at sector level s as follows:

$$\Psi_{s,\Delta} = \hat{\beta}_{s,0} + \hat{\beta}_{s,1} x_{1,\Delta} + \cdots + \hat{\beta}_{s,p} x_{p,\Delta}, \tag{4.26}$$

where $\mathbf{x}_\Delta = (x_{1,\Delta}, \ldots, x_{p,\Delta})'$ is the vector of stressed macroeconomic variables.

The difference between the stressed logit($\Psi_{s,\Delta}$) and the logit in t_0 is a key ingredient of the overall framework as shown below:

$$\Delta\zeta_s = \ln\left(\frac{\Psi_{s,\Delta}}{1 - \Psi_{s,\Delta}}\right) - \ln\left(\frac{\Psi_{s,t_0}}{1 - \Psi_{s,t_0}}\right), \tag{4.27}$$

where $\Delta\zeta_s$ is added to the initial logit as follows:

$$PD_{i,s,\Delta} = \frac{1}{1 + e^{-(\zeta_{i,s,t} + \Delta\zeta_s)}}. \tag{4.28}$$

The stressed default probability described in Eq. (4.28) is at the very heart of both the stress testing process and the risk integration framework. Moreover, the same idea is extendible to default probabilities estimated by use of other methods and models (e.g., probit).

It is worth mentioning that IRB default probabilities are calibrated on through-the-cycle outcomes. However, a point-in-time default probability is required to estimate provisions. For this reason, additional studies would be required to capture trend and cyclical components as, for example, in Carlehed and Petrov (2012). Furthermore, the method summarized in Eq. (4.28) may be characterized by the use of a principal component analysis or partial least squares technique (Krzanowski, 2000).

A different perspective is followed when a panel analysis is applied. In this case, macroeconomic variables are part of the core probability of default estimation process. The key drawback is that debtor-specific variables (e.g., credit bureau information and balance sheet ratios) usually play a predominant role compared with macroeconomic components.

The next section provides some indications on how to stress LGD.

4.4.2   Loss Given Default and Macroeconomic Variables

The LGD represents the percentage of the debtor's exposure that a bank expects to lose in the event of default. Many scenarios may be triggered after the latter event. The extremes are as follows. The bank recovers without any loss. On the contrary, the bank does not recover anything. A continuum of other occurrences lie between these extremes. The LGD (for a given facility i[4]) is estimated as a function of the ratio between the present value of net cash flows (i.e., recovery flows minus costs) and the exposure at the moment of default:

$$LGD_i = 1 - \frac{\sum_{t=1}^{T} PV(RE_{i,t}) - \sum_{t=1}^{T} PV(REC_{i,t})}{A_{i,t=def}}, \qquad (4.29)$$

where PV stands for present value, $RE_{i,t}$ are the recovery flows, $REC_{i,t}$ represents recovery costs over $t+1, \ldots, T$, and $A_{i,t=def}$ is the exposure at default. Both the estimation of net cash flows and the choice of the interest rate used to compute the present value affect the final assessment. In practice, the LGD is computed as an average of grouped debtor and facilities sharing some common features, such as collateral, loan to value, and so on. This aggregation serves the purpose of restricting the number of parameters to use in practice (i.e., LGD_{pt} represents the product type LGD).

In the last few years the concept of downturn LGD has been widely used by regulators to emphasize the need to incorporate the impact of the financial crisis in the LGD assessment. However, a downturn profile does not represent a macroeconomic framework for stressing LGDs but it can be a useful reference point.

All in all, a functional relationship similar to Eq. (4.25) may be used for LGD. The following equation captures the relationship between LGD and macroeconomic variables:

$$LGD_{pt,t} = \theta_{pt,0} + \theta_{pt,1}x_{1,t} + \cdots + \theta_{pt,p}x_{p,t} + \epsilon_{pt,t}, \qquad (4.30)$$

where the vector $\mathbf{x} = (x_1, \ldots, x_m)'$ represents macroeconomic variables. The regression coefficients are $\boldsymbol{\theta}_{pt} = (\theta_{pt,0}, \ldots, \theta_{pt,p})'$ and $\epsilon_{pt,t}$ is the error term. For stress testing purposes, the following holds:

$$LGD_{pt,\Delta} = \hat{\theta}_{pt,0} + \hat{\theta}_{pt,1}x_{1,\Delta} + \cdots + \hat{\theta}_{pt,p}x_{p,\Delta}. \qquad (4.31)$$

From a stress testing perspective, some banks use more sophisticated modeling techniques (e.g., the vector error-correction model described in Chapter 2). Nevertheless, it is worth noting that a critical issue arises when one is dealing with stress testing LGD. Indeed, the workout period usually covers several years. As a consequence, if a time series LGD allocates the overall result of the recovery process to the date of default, a misalignment problem emerges. Let us consider an extreme case where default occurs at time t and a single recovery flow occurs 3 years later. In line with stress testing need to project

---

4. The simplifying assumption of one customer, one facility holds throughout the book.

future occurrences, one should place LGD outcomes at time t. Then, one should rely on macroeconomic time series up to t (i.e., $t - k, t - k + 1, \ldots, t - 1$). In contrast, the recovery is affected by macroeconomic conditions along the interval $[t, t + 3]$. Such a peculiarity should be captured through the model.

In the next section the portfolio credit risk modeling and risk parameter stress testing are merged to highlight the key features of the risk integration framework proposed in the final two chapters.

4.5 PORTFOLIO CREDIT RISK STRESS TESTING

In Section 4.2 a wide family of portfolio models was presented and in Section 4.3 the regulatory RWAs were examined by our focusing on the IRB formula. Section 4.4 described how to stress the parameters feeding portfolio models and RWAs. It is now interesting to investigate how a portfolio stress testing may be structured by use of all the above components. In this regard, the next section highlights how to stress the regulatory RWAs. Then the subsequent section enters into the details of a more general portfolio stress testing that will be used in the final two chapters of the book as a base for the risk integration.

4.5.1 Stress Testing Risk-Weighted Assets

The RWA stress testing process may be seen from two broad standpoints. On the one hand, the standardized approach relies on fixed weights. Connecting these weights to macroeconomic drivers is a plausible exercise when a link exists with external rating. In all other cases, a one-to-one claim-category analysis is required. On the other hand, the IRB RWAs are a function of the credit parameters studied in the previous sections. Then one may use shocked risk inputs to estimate the stressed RWAs.

As described in Section 4.3, the standardized approach relies on a set of given weights. On this subject, a distinction arises between rating-driven and nonrating-related weights. When regulatory weights are a function of ratings, the method described in Section 4.4.1 may be used. In all other cases, assumptions need to be made on the basis of the nature of the claims. For example, a real estate market crisis may affect the loan-to-value ratio (i.e., the ratio of the loan amount and the value of the real estate collateral). Therefore a 35% weight should be applied on a reduced portion of the loan, with 100% being the weight to use for the unsecured part of the loan. A similar reasoning applies to claims secured by commercial real estate (i.e., portion of loans secured on which there is a 50% weight). All in all, credit risk mitigations may be affected by adverse macroeconomic conditions. The same applies to past due loans (i.e., nonperforming portfolio). On this subject, macroeconomic conditions may affect the coverage portion of specific provisions. Thus, the 100% weight may be replaced by a 150% weight if the provisions are less than 20% of the outstanding loan (see Section 4.3.1).

In the case of the IRB approach, credit risk parameters may be stressed according to the method described in Section 4.4. Example 4.9 shows the impact of shocked default probabilities and LGDs due to an adverse macroeconomic scenario.

Example 4.9 Stressed IRB RWAs

Let us consider the six-debtor portfolio described in Example 4.8. The equation estimated in Example 4.1 is used to assess the impact of the scenario described in Chapter 2. On this, the stressed scenario is such that UK GDP falls by 2.94% and UK inflation reduces by 0.28% (i.e., the stress testing inflation level becomes 0.23%) and no change is recorded in the short-term interest rate (r_t^{ST}). Table 4.15 summarizes the linear model coefficients, shocked macroeconomic variables values x_Δ, and the corresponding impact on the creditworthiness index.

TABLE 4.15 Default Probability Stress Testing Coefficients

Variable	Coefficient	x_Δ		
Intercept	0.04917			
GDP_{growth}	−0.94798	−2.94		
Δp_t	1.25060	0.23[a]		
r_t^{ST}	−3.29920			
ψ_{s,t_0}			5.83%	
$\ln\left(\frac{\psi_{s,t_0}}{1-\psi_{s,t_0}}\right)$			−2.78	
$\psi_{s,\Delta}$				7.99%
$\ln\left(\frac{\psi_{s,\Delta}}{1-\psi_{s,\Delta}}\right)$				−2.44

[a] Level of inflation.

Starting from Table 4.15, we apply a logit variation $\Delta\zeta_s = -2.44 + 2.78 = 0.34$ to each customer to obtain the stressed default probabilities (see Table 4.16). A 45% LGD (instead of the initial 40%) is then applied to assess the stressed RWAs. Therefore, an increase of $1.49 million (i.e., $8.34 million − $6.85 million) in the portfolio RWAs is outlined.

TABLE 4.16 Bank Rho Credit Risk-Weighted Assets ($ Millions)

Customer	Segment	PD t_0 (%)	ζ_{i,s,t_0}	$\Delta\zeta_s$	$\zeta_{i,s,t_0} + \Delta\zeta_s$	PD stressed (%)	RWAs
1	SME	0.50	−5.29	0.34	−4.95	0.70	0.95
2	SME	2.00	−3.89	0.34	−3.55	2.78	1.31
3	SME	3.00	−3.48	0.34	−3.14	4.16	1.42
4	Corporate	0.50	−5.29	0.34	−4.95	0.70	1.18
5	Corporate	2.00	−3.89	0.34	−3.55	2.78	1.66
6	Corporate	3.00	−3.48	0.34	−3.14	4.16	1.82
Total							8.34

PD, Probability of default; RWAs, risk-weighted assets; SME, small- and medium-sized enterprise.

The next section highlights some of the drawbacks of the portfolio modeling described above by introducing some of the key concepts to be used in Chapters 7 and 8.

4.5.2 Portfolio Credit Stress Testing

In the previous section stressed credit risk parameters were used to feed regulatory formula to derive stressed IRB RWAs. One of the main drawbacks of this approach is to rely on the infinitely granular portfolio assumption, which implies that no concentration risk affects the portfolio. Additionally, the IRB formula does not consider the impact of adverse conditions on default correlation (i.e., the correlation coefficient does not change under stress). The impact of these two elements may be highlighted through Example 4.10.

Example 4.10 How Copula Parameters Affect a Credit Loss Distribution

Let us start from Example 4.6, but instead of assuming customers are spread among 20 sectors, we now assume them to be distributed among five sectors. Additionally, it is worth highlighting how copula correlation may affect the loss distribution. For this reason, two alternatives are taken into account. Firstly, the analysis is conducted by use of a low correlation $\rho = 0.10$. Then a higher correlation $\rho = 0.50$ is used to generate the credit loss. As a result, Fig. 4.8 highlights the comparison of the loss distribution obtained by use of the three settings: (1) customers spread among 20 sectors and $\rho = 0.10$; (2) customers spread among five sectors and $\rho = 0.10$; (3) customers spread among five sectors and $\rho = 0.50$.

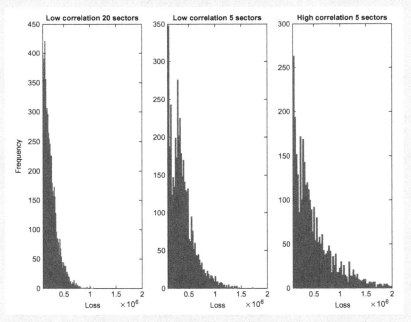

FIG. 4.8 Portfolio loss distribution computed by use of alternative assumptions on sector distribution and correlation.

Example 4.10 How Copula Parameters Affect a Credit Loss Distribution—cont'd

Table 4.17 highlights that the unexpected loss oscillates between 6.72% and 18.60% of the portfolio exposure depending on the sector distribution and copula correlation parameters. At the same time it is evident that none of these assumptions affect the IRB capital requirement. In other words, the IRB formula is not affected by the distribution among sectors or by the correlation among customer default.

TABLE 4.17 Unexpected Loss and Internal Ratings-Based Capital Requirement Under Different Assumptions

	20-sector $\rho = 0.10$	5-sector $\rho = 0.10$	5-sector $\rho = 0.5$
$UL_{VaR_{99.9\%}}$ (%)	6.72	11.84	18.03
$UL_{EL_{99.9\%}}$ (%)	7.73	12.63	18.60
$IRB_{cap.req.}$ (%)	7.74	7.74	7.74

A more comprehensive approach should be followed to incorporate changes in credit risk parameters linked to macroeconomic shocks and to take into account a more accurate estimate of default correlation. The key steps of a comprehensive credit stress testing portfolio modeling aligned with a bottom-up risk integration process are as follows:.

- **Simulate macroeconomic scenarios.** The first step is to simulate scenarios. A macroeconomic model in line with what was described in Chapter 2 may be used to generate coherent paths.
- **Estimate the impact of each scenario on credit risk parameters.** The second step is to derive shocked credit risk parameters. Eq. (4.28) may be used for default probabilities and Eq. (4.30) may be used for LGDs. In Chapter 5 additional evidence will be provided in terms of exposure projection.
- **Generate default events.** A framework that jointly considers macroeconomic scenarios and debtor-specific features is required. For this reason a copula approach may be used to generate defaults in line with what was described in Section 4.2.3.
- **Derive the loss distribution.** The last step of the process is to compute the credit portfolio loss as described in Eq. (4.11) by use of shocked credit risk parameters and a coherent default correlation structure.

The above framework is at the very heart of the risk integration process described in Chapter 7, where, apart from the credit risk, market, interest rate, and liquidity risks are taken into account by consideration of a comprehensive asset and liability structure.

4.6 SUMMARY

In this chapter the banking book was examined through the lens of portfolio credit risk modeling. A key distinction was made between expected and unexpected credit losses. Randomness was assumed to apply only to the probability of default. In this regard, a distinction was made between a structural (probit) and a logit approach. In the first case, the key idea was to identify the default on the basis of the comparison of asset and liability value. Default occurs when the asset value falls below the liability exposure. The CreditMetrics model was examined as an enhancement of the Merton-like structural model. In line with the intensity-of-default approach, a logit probability of default was used to feed a framework where default interdependencies were captured through a copula function. Aiming to assess the credit risk capital requirement, we investigated the standardized approach and the IRB method. This latter was derived as a special case of the more general credit portfolio modeling framework. Finally, the relation between risk parameters and macroeconomic variables was studied. A portfolio stress testing method was outlined by our pointing out the role of default probability, LGD, exposure, and copula parameters in assessing an overall portfolio credit risk.

SUGGESTIONS FOR FURTHER READING

Starting from the seminal documents produced on the eve of the credit portfolio management, the key ideas behind the methods still in use in current risk practice are described in CreditMetrics (1997), CSFB (1997), and Wilson (1997). Duffie and Singleton (2003) explore the key items related to credit pricing, and Bielecki and Rutkowski (2001) present a very detailed and mathematically comprehensive description of credit risk techniques.

From a regulatory perspective, BIS (2006) is a milestone for understanding all the details behind the capital requirement assessment. The most recent accord, Basel III (BIS, 2011), relies on the credit risk architecture defined under Basel II.

Finally, a series of research articles explore stress testing from a credit risk standpoint. Among others, Drehmann et al. (2010) propose a very interesting integrated stress testing framework, while Castren et al. (2010) use a global vector autoregression model to analyze euro area corporate default probabilities.

APPENDIX A: DEFAULT PROBABILITY ESTIMATION VIA LOGIT REGRESSION

In linear regression, data consist of pairs of observations. Numerical (continuous measurement) variables are used to predict a numerical response variable. On this subject, the linear model consists of two parts: a structure on the means and an error component. The mean response is assumed to be a linear function of the explanatory (predictor) variable. The error structure in the model attempts to describe how individual measurements vary around the mean value. In this regard, responses are assumed to vary around the mean according to a normal distribution with variance σ^2. In a univariate setting, the linear model may be expressed as follows:

$$\mathbb{E}(y_r|z_r) = \beta_0 + \beta_1 z_r, \tag{4.32}$$

where r stands for a generic observation, and errors are i.i.d. $\epsilon_r \sim N(0, \sigma^2)$.

By contrast, the aim of default probability models is to fit binary responses ($y_r = 1$ in the case of default, $y_r = 0$ in the case of nondefault). This can be represented in terms of probability as follows: $P(y_r = 1) = \pi_r$, $P(y_r = 0) = 1 - \pi_r$. Therefore the structure of the means is

$$\mathbb{E}(y_r|z_r) = \beta_0 + \beta_1 z_r = \pi_r, \tag{4.33}$$

and the structure of errors is such that the variance is a function of π_r:

$$\sigma^2(y_r) = \pi_r(1 - \pi_r), \tag{4.34}$$

which violates the linear regression assumption of constant σ^2.

According to the above, when the response variable is binary, the expected response is more appropriately modeled by some curved relationship with the predictor variable. One such curved relationship is given by the logistic model

$$\mathbb{E}(y_r|z_r) = \pi_r = \frac{e^{\beta_0+\beta_1 z_r}}{1 + e^{\beta_0+\beta_1 z_r}} = \frac{1}{1 + e^{-(\beta_0+\beta_1 z_r)}}. \tag{4.35}$$

This function is particularly useful for several reasons. Firstly, it is bounded between 0 and 1. This eliminates the possibility of getting predictions outside the [0,1] interval. Secondly, there is a linear model hidden in the function that can be revealed with a proper transformation of the response. Finally, the sign associated with the coefficient β_1 indicates the direction of the curve. A positive value for β_1 indicates an increasing function, while a negative value indicates a decreasing function.

With regard to the above linear model hidden within the logistic model, the logit transformation uncovers this relationship. On this, the natural logarithm of the ratio of π_r to $(1 - \pi_r)$ gives a linear model in z_r as follows:

$$\ln\left(\frac{\pi_r}{1 - \pi_r}\right) = \beta_0 + \beta_1 z_r. \tag{4.36}$$

Parameters are estimated by maximization of the following likelihood function:

$$L(\beta_0, \beta_1|Data) = \prod_{r=1}^{n_r} \pi_r^{y_r}(1 - \pi_r)^{1-y_r} \tag{4.37}$$

where, by substituting $\pi_r = \frac{e^{\beta_0+\beta_1 z_r}}{1+e^{\beta_0+\beta_1 z_r}}$ and $(1 - \pi_r) = \frac{1}{1+e^{\beta_0+\beta_1 z_r}}$ into the above equation, we obtain

$$L(\beta_0, \beta_1|Data) = \prod_{r=1}^{n_r}\left(\frac{e^{\beta_0+\beta_1 z_r}}{1 + e^{\beta_0+\beta_1 z_r}}\right)^{y_r}\left(\frac{1}{1 + e^{\beta_0+\beta_1 z_r}}\right)^{1-y_r} = \prod_{r=1}^{n_r}\frac{(e^{\beta_0+\beta_1 z_r})^{y_r}}{1 + e^{\beta_0+\beta_1 z_r}}. \tag{4.38}$$

It is usually easier to estimate parameters by use of the log-likelihood as follows:

$$\log[L(\beta_0, \beta_1 | Data)] = \sum_{r=1}^{n_r} y_r(\beta_0 + \beta_1 z_r) - \sum_{r=1}^{n_r} \log\left(1 + e^{\beta_0 + \beta_1 z_r}\right), \quad (4.39)$$

where the goal is to choose β_0 and β_1 so as to maximize the log-likelihood. Similarly to ordinary least squares, there will be two equations that must be solved for two unknowns. Unlike ordinary least squares, the two equations will not be linear and so must be solved by iteration (start with initial values for β_0 and β_1, evaluate the log-likelihood, choose a new value for β_0 or β_1 that reduces the log-likelihood, and repeat the process until the log-likelihood does not change).

The above model can be easily extended to the multivariate case as shown in Exercise 4.1.

APPENDIX B: THE FORWARD SEARCH FOR ELLIPTICAL COPULAS

As examined in the previous chapters, atypical observations may affect the overall model estimation. The forward search may be used to detect outlying units (Bellini, 2016). On this subject, one first needs a method to estimate copula parameters. Then a measure of unit closeness is required. Finally, the forward search is performed.

The starting point for the estimation process is the time series vector $\mathbf{y}_t = (y_{1,t}, \ldots, y_{S,t})'$ of observations. In terms of notation, S stands for the number of time series under analysis (e.g., number of sectors under investigation) and $t = 1, \ldots, T$ is a discrete time. In the literature, several methods have been proposed to estimate copula parameters (Cherubini et al., 2004). In this regard, the canonical maximum likelihood (CML) is the most commonly adopted approach. Firstly, it avoids a priori assumptions on the distributional form of the marginals. Secondly, it is easy to implement even for high-dimensional copulas. In the case of elliptical copulas, the CML process may be summarized as follows:

- **Transformation of the time series in uniform variables.** The first step of the CML process is to transform the initial dataset $\mathbf{y}_t = (y_{1,t}, \ldots, y_{S,t})'$ into uniform variables. This is done by use of the empirical marginal distribution as detailed below:

$$\hat{u}_{s,t} = \frac{1}{T} \Sigma_{q=1}^T \mathbb{1}_{y_{s,q} \leq y_{s,t}}, \quad (4.40)$$

where

$$\mathbb{1}_{y_{s,q} \leq y_{s,t}} = \begin{cases} 1 & \text{for} \quad y_{s,q} \leq y_{s,t}, \\ 0 & \text{for} \quad y_{s,q} > y_{s,t}. \end{cases} \quad (4.41)$$

- **Copula parameter estimation.** The second step of the process is to estimate copula parameters (i.e., correlation matrix (ρ) for the normal copula, and ρ together with degrees of freedom (ϑ) for the Student T copula). In this regard, each element of ρ is estimated by use of

$$\tau(y_s, y_r) = \frac{2}{\pi} \arcsin(\rho_{s,r}), \tag{4.42}$$

where $\tau(y_s, y_r)$ and $\rho_{s,r}$ indicate the Kendall τ and the Pearson linear correlation coefficient, respectively. The estimator of $\rho_{s,r}$, $\sin[\frac{\pi}{2} \hat{\tau}(y_s, y_r)]$, inherits the robustness of the Kendall τ estimator. Additionally, it is an efficient estimator for both elliptical and nonelliptical distributions (Embrechts et al., 2001). However, there is no guarantee that the empirical Kendall τ transformation matrix will be positive definite. When it is not positive definite, the eigenvalue method of Rousseeuw and Molenberghs (1993) may be used to perform an adjustment.

For the Student T copula, Mashal and Zeevi (2002) proposed the following algorithm to obtain the degrees of freedom ϑ:

- The initial dataset is transformed into a set of uniform variates $\hat{u}$ by use of the empirical marginal transformations described above.
- The correlation matrix is estimated on the time series under analysis.
- The following log-likelihood density function is maximized to estimate ϑ on the basis of the CML approach:

$$\hat{\vartheta}_{CML} = \underset{\vartheta}{\mathrm{argmax}} \sum_{t=1}^{T} \ln\left[c_{Student}(\hat{u}_{1,t}, \ldots, \hat{u}_{S,t} | \hat{\rho}, \vartheta) \right]. \tag{4.43}$$

The forward search needs a measure of unit closeness. Therefore, starting from $\hat{u}_{s,t}$ of Eq. (4.40), the following inverse marginal is computed:

$$\tilde{u}_{s,t} = F^{-1}(\hat{u}_{s,t}), \tag{4.44}$$

where for the Gaussian copula $F^{-1}(\cdot)$ is the inverse of the standard univariate normal CDF $\Phi^{-1}(\cdot)$. For the Student T copula, the standard univariate CDF $t_{\vartheta}^{-1}(\cdot)$ is used. Considering that $\hat{u}_{s,t}$ is the rank of $y_{s,t}$ divided by T, $\hat{u}_{s,t}$ is set to $\frac{rank-0.5}{T}$ to prevent $F^{-1}(\hat{u}_{s,t})$ from reaching its extreme limits. Therefore the measure to use to cause the forward search to progress relies on $\hat{\rho}$ and $\tilde{u}_{s,t}$ as detailed below:

$$d_t(m^*) = \tilde{\mathbf{u}}_t' \hat{\rho}^{-1} \tilde{\mathbf{u}}_t. \tag{4.45}$$

In line with the forward search analysis shown in Chapters 2 and 3, the monitoring of the search is conducted by use of the minimum distance of units outside the subset:

$$d_{tmin}^*(m^*) = \min[d_t(m^*)] \quad t \notin S_*^{(m)}. \tag{4.46}$$

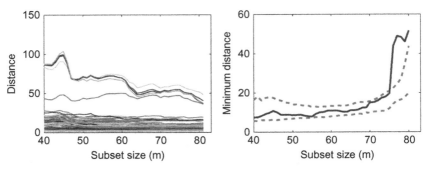

FIG. 4.9 Forward distance plots and envelopes in a contaminated setting.

The forward search procedure is perfectly aligned with what was shown in previous chapters. In what follows, a brief example shows its main features (Bellini, 2010). Let us simulate a normal copula with $S = 6$ time series of length $T = 81$. The correlation parameter is set at 0.01 for all pairs of variables. When the time series are not contaminated, no outlier is detected. Then a contamination scheme is applied. For this reason, units 20 to 24 are exogenously corrupted. On this subject, the left-hand side of Fig. 4.9 highlights these five units as a separate cluster. The right-hand side shows that $d^*_{tmin}(m^*)$ dramatically increases when the contaminated observations enter the subset. A superimposition process confirms that all five units may be considered as atypical.

EXERCISES

Exercise 4.1 Let us consider the database Chap4PDlogit.xlsx where 457 debtors are recorded according to their default status: 1 default, 0 nondefault. The firm-specific explanatory variables are liquidity ratio (*Liquid*), number of credit delay days (*DDcredit*), return on assets (*ROA*), and the ratio representing the bank's credit line use compared with the use over the entire banking system (*Use*).

Implement the R code to develop a logit model by comparing the effectiveness of using the variable *Liquid* only against the use of all explanatory variables. Thus, show a chart representing true positive, true negative, false positive, and false negative estimates as well as the receiver operating characteristic (ROC) area curve.

Exercise 4.2 Apply the model described above to estimate the default probabilities for the portfolio Chap4PDRWA.xlsx. Then compute the IRB RWAs with use of what was described in Section 4.3.2.

Exercise 4.3 Use the model estimated on the UK economy to stress the model parameters of Exercise 4.2. Then estimate the corresponding stressed RWAs.

Solutions are available at www.tizianobellini.com.

REFERENCES

Altman, E., 1968. Financial ratios, discriminant analysis and the prediction of corporate bankruptcy. J. Finance 23, 589–609.

Atkinson, A.C., Riani, M., Cerioli, A., 2004. Exploring Multivariate Data with the Forward Search. Springer, New York.

Bellini, T., 2010. Detecting atypical observations in financial data: the forward search for elliptical copulas. Adv. Data Anal. Classif. 4, 287–299.

Bellini, T., 2016. The forward search interactive outlier detection in cointegrated VAR analysis. Adv. Data Anal. Classif. 10, 351–373.

Bielecki, T., Rutkowski, M., 2001. Credit Risk: Modeling, Valuation and Hedging. Springer, Berlin.

BIS, 2006. Basel II International Convergence of Capital Measurement and Capital Standards: A Revised Framework. Bank for International Settlements, Basel.

BIS, 2011. Basel III: A global regulatory framework for more resilient banks and banking systems. Bank for International Settlements, Basel.

BIS, 2015. Consultative document: Revisions to the standardized approach for credit risk. Bank for International Settlements, Basel.

Carlehed, M., Petrov, A., 2012. A methodology for point-in-time-through-the-cycle probability of default decomposition in risk classification systems. J. Risk Model Validat. 6 (3), 3–25.

Castren, O., Dees, S., Zaher, F., 2010. Stress-testing euro area corporate default probabilities using a global macroeconomic model. J. Financ. Stab. 6, 64–74.

Cherubini, U., Luciano, E., Vecchiato, W., 2004. Copulas for Finance. Wiley, Chichester.

CreditMetrics, 1997. Creditmetrics. www.creditmetrics.com.

Credit Suisse First Boston, 1997. Creditrisk+: A credit risk management framework. Credit Suisse First Boston, New York.

Drehmann, M., Stringa, M., Sorensen, S., 2010. The integrated impact of credit and interest rate risk on banks: A dynamic framework and stress testing application. J. Bank. Finance 34, 713–729.

Duffie, D., Singleton, K.J., 2003. Credit Risk Pricing, Measurement and Management. Princeton University Press, Princeton.

Embrechts, P., Lindskog, F., McNeil, A., 2001. Modelling dependence with copulas and applications to risk management. Department of Mathematics, ETH Zurich.

Gordy, M., 2003. A risk-factor foundation for risk-based capital rules. J. Financ. Intermed. 12, 199–232.

Koyluoglu, H., Hickman, A., 1998. Reconcilable differences. Risk 11 (10), 56–62.

Krzanowski, W.J., 2000. Principles of Multivariate Analysis. Oxford University Press, Oxford.

Mashal, R., Zeevi, A., 2002. Beyond correlation: Extreme co-movements between financial assets. Columbia Graduate School of Business.

Merton, R., 1974. On the pricing of corporate debt: the risk structure of interest rates. J. Finance 29 (2), 449–470.

Rousseeuw, P.J., Molenberghs, G., 1993. Transformation of non positive semidefinite correlation matrices. Commun. Stat. Theory Methods 22, 965–984.

Schonbucher, P., Schubert, D., 2001. Copula dependent default risk in intensity models. University of Bonn.

Sklar, A., 1959. Fonctions de repartition a n dimensions et leur marges. Publications de l'Institut de Statistique de l'Universite' de Paris 106, 1039–1061.

Vasicek, O., 2002. Loan portfolio value. Risk 15 (2), 160–162.

Wilson, T., 1997. Portfolio credit risk (i). Risk 10, 111–117.

Chapter 5

Balance Sheet, and Profit and Loss Stress Testing Projections

Chapter Outline

This chapter describes how a stress scenario affects a bank's balance sheet and its profitability. Operational risk is also explored in relation to profit and loss projections.

Given the role played by a credit portfolio in a commercial bank, a credit life cycle process representation is a useful starting point for balance sheet projections. Flows between performing and nonperforming portfolios are investigated as a core element of the stress testing mechanism. This is the starting point to assess credit losses and outline a bank's overall risk profile. In this regard, the trading book, other assets, and liabilities are also studied to link the balance sheet and the profit and loss statements.

Focusing on the profit and loss, the preprovisioning net revenue (PPNR) is represented as the sum of net interest income (NII), noninterest revenue (NIR), and noninterest expenses (NIEs). Then, a connection with the credit portfolio is described thorough exploration of loan impairment charges (LICs). A key

Stress Testing and Risk Integration in Banks. http://dx.doi.org/10.1016/B978-0-12-803590-0.00005-9

distinction arises among collective provisions, specific provisions, and write-offs. Net profit is explored as the final ring connecting the balance sheet and profit and loss chain.

Finally, conduct and operational risks are described. The focus is on the projection of losses and their impact on risk-weighted assets. Three regulatory approaches are briefly illustrated by our showing the contribution of operational risk as a core component of the pillar 1 capital requirement.

KEY ABBREVIATIONS AND SYMBOLS

$A_{gross,t}$	performing asset at time t, gross from provisions
$A_{gross,NP,t}$	nonperforming asset at time t
$AM_{h,\Delta}$	amortizing balance during period h under macroeconomic scenario Δ
$BG_{h,\Delta}$	balance growth during period h and scenario Δ
CP_t	collective provisioning at time t
$CU_{h,\Delta}$	credit cure flow during period h and scenario Δ
$DF_{h,\Delta}$	default flow during period h and scenario Δ
L_t	liability balance at time t
$LGD_{def,i,t}$	loss given default for defaulted customer i at time t
$LIC_{h,\Delta}$	loan impairment charge over the period h and scenario Δ
$NIE_{h,\Delta}$	noninterest expenses over period h and scenario Δ
$NII_{h,\Delta}$	net interest income over the period h under macroeconomic scenario Δ
$NIR_{h,\Delta}$	noninterest revenue over period h and scenario Δ
$OC_{i,t}$	operating cost for customer i at time t
PCU_t	probability of cure at time t
$PPNR_{h,\Delta}$	preprovisioning net revenue over the period h and scenario Δ
$RF_{i,t}$	recovery flow for customer i at time t
SP_t	specific provisioning at time t
TB_t	trading book balance at time t
$UR_{i,t}$	utilization rate for customer i at time t
$WO_{h,\Delta}$	write-off flow during period h and scenario Δ

5.1 INTRODUCTION

Starting from Chapter 2, we depicted a regulatory stress scenario and introduced tools for its enrichment. An overall balance sheet representation characterized Chapter 3, where Bank Alpha was introduced as a stylized commercial bank. The focus was on margin, value at risk, and liquidity analysis. A deep dive into the credit portfolio distinguished Chapter 4. Now all the instruments required to project a balance sheet and assess profit and loss impacts are on the shelf.

Stakeholders are usually keen to understand the projection phase of the stress testing process. In this regard, one of the main issues a bank needs to face when implementing a stress testing framework is to design the interactions between the balance sheet and profit and loss. On this, the credit process contains a series of peculiarities. For this reason, a detailed credit life cycle study is shown in Section 5.2. Credit deterioration is one of the major topics that regulators are keen to monitor under a stressed scenario. Therefore collective and specific provisions need to be examined as part of the overall architecture. All in all, a comprehensive investigation of the whole balance sheet evolution is required and rules to project the banking book, trading book, and liabilities are vital for the entire stress testing mechanism.

Regulators usually supply a high-level description of the process to be followed for a stress testing exercise. This chapter attempts to fill the gap between general principle statements and a more comprehensive detailed representation needed by a bank for a practical implementation.

Section 5.3 pursues the goal of building a bridge connecting the balance sheet and the profit and loss. As a starting point, one needs to highlight that assets and liabilities generate interest. The balance of interest revenue and interest expenses results in the NII. NIR due to commission and fees and NIEs related to operational activities are added to the NII to compute the PPNR. The credit assessment completes the profit and loss scrutiny by pointing out the contribution of LICs to a bank's net profit (loss).

As a part of the pillar 1 capital requirements, Section 5.4 highlights operational risk connections with the entire profit and loss analysis. The focus is on the key links between operational risk and the overall stress testing framework. Adequate references are supplied for a more comprehensive discussion of operational risk.

In terms of the toolkit, this chapter follows an approach slightly different from that in previous chapters. In particular, given the complexity of the credit process, the credit life cycle work flow is the first tool the reader is required to become familiar with. Diagrams together with (linear) equations summarizing how to account for interdependencies are widely used throughout the chapter. Accounting and statistical techniques to link balance sheet items to external macroeconomic variables complete the set of instruments the reader needs to be armed with to master the stress testing projection process detailed in the following sections.

5.2 BALANCE SHEET PROJECTION

A distinction is usually made between the projection of banking book instruments compared with other balance sheet elements. Within the banking book, the credit deterioration process is crucial to pinpoint macroeconomic impacts on a bank's risk profile. For this reason, a credit life cycle analysis kicks off the balance sheet projection pipeline in Section 5.2.1. The evolution of the trading book, other assets, and liabilities is outlined in the subsequent sections.

5.2.1 Credit Life Cycle

As a premise, the banking book encompasses all financial instruments that are not actively traded and are meant to be held until their maturity. The vast bulk of the banking book is commonly represented by credits. In this regard, regulators do not usually detail all microstructure interactions between performing and nonperforming portfolios. Therefore a bank is required to outline its own process and implement a comprehensive framework. In what follows, a typical credit life cycle process is outlined. The description of good practice to project credit stocks and flows is the crucial purpose of this section.

A credit life cycle process may be summarized as detailed in Fig. 5.1. The portfolio is split into performing and nonperforming credits (i.e., deteriorated including defaulted). Additionally, a distinction is made between asset exposure indicated with a plus sign and provisions denoted with a minus sign. This graphical structure is aligned with the accounting practice to record credits and provision stocks in separate accounts. Nonetheless, one needs to bear in mind that credits are usually reported net of provisions in the financial statement. It is

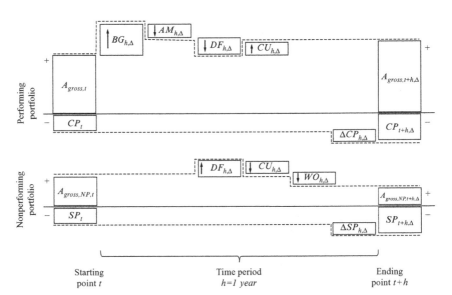

FIG. 5.1 Credit life cycle dynamics over a 1-year time horizon. Performing and nonperforming portfolios.

worth noting that a bank may have additional segments and other subportfolios that should be included in the credit life cycle picture. Nonetheless, Fig. 5.1 may be considered as representative of a typical commercial (wholesale and retail) banking process. This structure will be used as a map throughout this chapter.

Starting from time t, Fig. 5.1 shows the key elements required to project a bank's balance sheet over the period h. In this regard, a useful distinction is made between stock and flow components. In Fig. 5.1, asset exposure (A), collective provisions (CP), and specific provisions (SP) are stock variables. All others are flows. The latter have the subscript h, Δ encompassing the following two assumptions. Firstly, h is the period of the projection. It is set equal to 1 year in line with the usual accounting and regulatory exercise. For a multiperiod stress test, subsequent projections are required. Secondly, Δ indicates the macroeconomic scenario on which the exercise relies. One needs to bear in mind that all flows occur contemporaneously. As a consequence, the representation highlighting the balance sheet growth ($BG_{h,\Delta}$) preceding the amortization ($AM_{h,\Delta}$) and the default flow ($DF_{h,\Delta}$) does not subsume $BG_{h,\Delta} \rightarrow AM_{h,\Delta} \rightarrow DF_{h,\Delta}$. These flows are generated during the entire interval h without solution of continuity. In what follows, all stock and flow components are analyzed and their role is explained as part of the overall framework.

Let us start the analysis from the performing asset ($A_{gross,t}$). The subscript *gross* indicates that the amount of credit is gross from collective provisions (CP_t). As we move to the right-hand side, a balance growth ($BG_{h,\Delta}$) is shown. Usually $BG_{h,\Delta}$ increases the balance. Nonetheless, when a business is restructured or the bank exits some markets, a reduction may occur. It is worth mentioning that for stress testing purposes the FRB (2015) and the BOE (2015) (among others) use a dynamic balance sheet. In contrast, the EBA (2016) relies on a static balance sheet by assuming that no changes in the exposure will occur over the stress testing period. It implies that all instruments reaching their maturity within this interval are replaced by other instruments with the same financial and risk characteristics.

In line with a dynamic approach, the FRB (2015) developed a model for Comprehensive Capital Analysis and Review purposes to project industry-wide total assets, loans, securities, and total liabilities. This model represents a balance sheet evolution coherent with past crises. In more detail, the model does not project individual asset categories but tends to capture the shift in bank assets that occurs naturally under stress. Hence one can easily translate total growth rates into firm-specific projections by assuming product mixes and market share are constant over time. Banks are solicited to develop frameworks for projecting their exposures in line with macroeconomic scenarios and their historical evolution.

As anticipated above, amortization or expiry at maturity ($AM_{h,\Delta}$) flows cause a reduction in exposures as indicated in the third box in the top panel in Fig. 5.1.

Flows from the performing portfolio to the nonperforming portfolio (and vice versa) qualify the credit process. More specifically, default flows ($DF_{h,\Delta}$) decrease the performing portfolio by correspondingly increasing the nonperforming balance. In contrast, credit cure ($CU_{h,\Delta}$) decreases the nonperforming portfolio by increasing the performing portfolio.

The end-of-period gross exposure ($A_{gross,t+h,\Delta}$) is obtained as the sum of the initial stock and all the above-mentioned flows.

In the lower part of the performing portfolio picture, CP_t mirrors the gross outstanding balance. CP_t acts as a buffer against expected losses. A flow of new collective provisions ($\Delta CP_{h,\Delta}$) occurs during the interval h based on the performing portfolio renewed shape (i.e., new mix of credits, changed credit risk, and so on). The end-of-period collective provision stock is obtained as the sum of CP_t and $\Delta CP_{h,\Delta}$.

The lower panel in Fig. 5.1 focuses on the nonperforming portfolio. In this case, the starting point is the defaulted exposure $A_{gross,NP,t}$. The sum of $A_{gross,t}$ and $A_{gross,NP,t}$ represents the overall credit balance of a bank at time t. A series of flows affect the evolution of $A_{gross,NP}$ during the interval h. As stated earlier, the default flow $DF_{h,\Delta}$ increases and $CU_{h,\Delta}$ reduces defaulted exposures. At the end of the recovery process, write-offs ($WO_{h,\Delta}$) are recorded when a misalignment between estimated and actual losses occurs. The end-of-period net outstanding balance is in essence the sum of the initial stock of nonperforming credits and the flows described above.

Finally, specific provision (SP) movements are summarized in terms of a variation due to a modified nonperforming portfolio mix ($\Delta SP_{h,\Delta}$). It is useful to note that the static balance sheet assumption informing the EBA (2016) stress testing implies a bipartition of $\Delta SP_{h,\Delta}$. In more detail, $\Delta SP_{h,\Delta}^{OLD}$ stands for the increase of specific provisions due to changes in the stock of provisions at the beginning of the period (i.e., SP_t). In contrast, $\Delta SP_{h,\Delta}^{NEW}$ is related to the default flow $DF_{h,\Delta}$. As an additional note, the European Banking Authority static balance sheet assumption also implies not taking into account cures and write-offs. Their impact in terms of profit and loss needs to be captured in the loss given default (LGD) estimate. It is worth mentioning that even in the case of a static balance sheet, flows from the performing to the nonperforming portfolio need to be considered. The same applies to the dynamic affecting nonperforming stock, collective provisions, and specific provisions. Therefore the framework introduced above is a paradigm for both a dynamic and a static balance sheet framework.

All in all, the performing portfolio evolution can be summarized as follows:

$$A_{gross,t+h,\Delta} = A_{gross,t} + BG_{h,\Delta} - AM_{h,\Delta} - DF_{h,\Delta} + CU_{h,\Delta}. \quad (5.1)$$

The collective provisions' stock evolution may be summarized as follows:

$$CP_{t+h,\Delta} = CP_t + \Delta CP_{h,\Delta}. \tag{5.2}$$

For the defaulted portfolio, the following holds

$$A_{gross,NP,t+h,\Delta} = A_{gross,NP,t} + DF_{h,\Delta} - CU_{h,\Delta} - WO_{h,\Delta}. \tag{5.3}$$

Finally, the specific provisions' dynamic can be represented in the following way:

$$SP_{t+h,\Delta} = SP_t + \Delta SP_{h,\Delta}. \tag{5.4}$$

The next section details how to project the performing portfolio. Afterward, the focus will move to the nonperforming portfolio.

5.2.2 Performing Portfolio Projection

As a first step in the analysis of the performing portfolio, a distinction between exposure and provisions is required. In this regard, Fig. 5.1 highlights that balance sheet growth, amortization, default flow, and cure characterize the exposure side (i.e., top panel in Fig. 5.1). The latter two also enter into the provision stock evolution in conjunction with $\Delta CP_{h,\Delta}$.

In what follows, a more detailed description of each component mechanism is shown and hints for the estimation process to be followed are supplied. The focus is on a 1-year period. Nevertheless, a rolling process may be applied. Along this line, no explicit reference will be highlighted on a multiperiod environment if it is not specifically needed for a comprehensive explanation.

- **Balance sheet growth** $(BG_{h,\Delta})$. Balance sheet exposure (A_t) would need to be linked to macroeconomic variables. A simple and intuitive way to pursue this goal is to regress each specific asset type k (e.g., a given facility or product type), against macroeconomic variables. The following equation summarizes the relationship:

$$A_{k,t} = \beta_{k,0} + \beta_{k,1} x_t + \cdots + \beta_{k,p} x_{p,t} + \epsilon_{k,t}, \tag{5.5}$$

where $\boldsymbol{\beta}_k = (\beta_{k,1}, \ldots, \beta_{k,p})'$ is the vector of coefficients fitted against the vector of macroeconomic variables $\mathbf{x}_t = (x_{1,t}, \ldots, x_{p,t})'$. Errors are i.i.d. normally distributed.

A more sophisticated method should use the vector autoregression or vector error-correction models described in Chapter 2. Furthermore, principal component analysis (Krzanowski, 2000) or partial least squares should be used. This would be convenient when the number of macroeconomic variables is high and all of them are significant to explain balance sheet evolution. In all cases, once the model has been estimated, the exposure growth can easily be assessed. For a given scenario $\mathbf{x}_\Delta$, the following holds:

$$BG_{k,h,\Delta} = \hat{A}_{k,\Delta} - A_{k,t} = \hat{\beta}_{k,0} + \hat{\beta}_{k,1}x_{1,\Delta} + \cdots + \hat{\beta}_{k,p}x_{p,\Delta} - A_{k,t}. \quad (5.6)$$

where $\hat{A}_{k,\Delta}$ is the exposure under stress and $A_{k,t}$ is the actual exposure.

As pointed out in Section 5.2 and highlighted in Chapter 4, the concept of asset exposure is critical in the Basel II Accord framework. On this subject, let us consider, as an example, the case where a bank committed $1 million (i.e., committed balance CB_t) to a customer whose utilization is $0.6 million (i.e., utilized balance UB_t). In this regard, the Basel II Accord requires assessment of the portion of the undrawn balance on which a bank may be exposed (within a 1-year period). This undrawn portion of the commitment is usually recorded among off-balance sheet items. Credit conversion factor are then applied to capture their potential future financial exposure. The following equation summarizes the modeling components that should be taken into account. Exposure amount at time $t + h$ conditional on the committed and the utilized balance at time t may be represented as follows:

$$A_{t+h|t} = UB_t + UR_{t+h|t}(CB_t - UB_t), \quad (5.7)$$

where $UR_{t+h|t}$ is the utilization rate conditional on t. The utilization rate needs to be applied to the gap existing between the committed and the utilized balance at time t. As a consequence, the following holds:

$$\frac{A_{t+h|t}}{UB_t} = \frac{[UB_t + UR_{t+h|t}(CB_t - UB_t)]}{UB_t} \quad (5.8)$$
$$= 1 + \left[UR_{t+h|t}\left(\frac{CB_t}{UB_t} - 1\right)\right] > 1,$$

where, under adverse scenarios, the utilization rate is expected to increase. Hence a comprehensive modeling needs to be put in place. UB_t, CB_t, and UR_t may be regressed against macroeconomic variables as detailed in Eq. (5.5). The notation A_{t+h} will continue to be considered in a broad sense in what follows (including exposure at default à la Basel II). The reader may use a simple idea of assets as exposure. No additional effort is required to distinguish between on-balance sheet and off-balance sheet items.

- **Amortization and expiry at maturity** $(AM_{h,\Delta})$. Flows causing a reduction in exposure may be caused by both the amortization scheme and expiry at maturity. These flows are represented through balance sheet growth in a dynamic setting. In contrast, in a static balance sheet framework, a replacement of $AM_{h,\Delta}$ is required. In this case, financial instruments are rolled over through operations with the same characteristics and risk profile.

- **Default flow** ($DF_{h,\Delta}$). Let us assume that a bank has a (1-year) probability of default for each asset exposure (i.e., $A_{i,t}$). According to what was described in Chapter 4, stressed default probabilities are computed ($PD_{i,\Delta}$). As a consequence, the default flow conditional on the macroeconomic scenario on which $PD_{i,\Delta}$ relies may be estimated as follows:

$$DF_{h,\Delta} = \sum_{i=1}^{n} A_{i,t} PD_{i,\Delta}, \tag{5.9}$$

  where n stands for the number of credits out of the N assets of the overall balance sheet. Eq. (5.9) does not capture concentration risk or other portfolio-specific features. In this regard, one should use Monte Carlo simulations to identify customers defaulting over the interval h. Nonetheless, given the complexity of a simulation process, banks rarely adopt this solution to assess the default flows for balance sheet projections. Eq. (5.9) is commonly adopted in practice.
- **Credit cured** ($CU_{h,\Delta}$). A bank runs a recovery procedure after default occurrence. This process may end up with a full, partial, or null recovery. In some circumstances, recovery cost may even exceed the amount recovered. It is common practice to indicate as cured the (fully) recovered credits. In what follows, the cure flow is represented as

$$CU_{h,\Delta} = \sum_{i_{def}=1}^{n_{def}} A_{i_{def},t} PCU_{i_{def},\Delta}, \tag{5.10}$$

  where $A_{i_{def},t}$ stands for a defaulted asset exposure in the initial period of analysis, while $PCU_{i_{def},\Delta}$ is the probability of cure, given a macroeconomic scenario.
- **Collective provisions** ($CP_{t+h,\Delta}$). Collective provisions operate as a buffer against expected losses. Despite the intuitive idea, provisions may be estimated according to alternative accounting principles. As an example, in response to the 2007–09 financial crisis, the International Accounting Standard Board proposed new standards. IFRS 9 focuses on expected losses. This principle draws a distinction between 1-year and lifetime expected losses. The concept of a significant increase in credit risk is used as a trigger for the 1-year or lifetime computation. The most updated US GAAPs refer to a lifetime principle for all credits. An easy way to overcome accounting specificities is to start from an expected loss equation based on the product of exposure at default, probability of default, and LGD. Then an accounting-specific adjustment is applied as detailed below:

$$CP_{t+h,\Delta} = \left(\sum_{i=1}^{n} A_{i,t} PD_{i,\Delta} LGD_{i,\Delta} \right) ADJ_h, \tag{5.11}$$

  where ADJ_h stands for the accounting-specific adjustment. As a related flow, the variation of collective provisions ($\Delta CP_{h,\Delta} = CP_{t+h,\Delta} - CP_t$) will be analyzed as part of the profit and loss estimation.

The next section focuses on the key elements characterizing the nonperforming portfolio evolution.

5.2.3 Nonperforming Portfolio Projection

In what follows, credit cure and specific provisions are investigated from the nonperforming portfolio standpoint. The key steps of the recovery process are examined accordingly.

- **Specific provisions** ($SP_{h,\Delta}$). When a default occurs, a bank activates a recovery process and a specific provision is recorded. If collective provisions aim to create a generic buffer against the overall portfolio losses, specific provisions focus on an individual assessment of defaulted credits. As detailed in Eq. (5.4), $SP_{h,\Delta}$ depends on the initial stock SP_t and a flow $\Delta SP_{h,\Delta}$. When a new default flow occurs, a bank increases its specific provisions in line with its recovery expectations. On this subject, an internally estimated LGD is used to assess expected losses as detailed below:

$$\Delta SP_{h,\Delta} = DF_{h,\Delta}LGD_{\Delta}, \tag{5.12}$$

where LGD_{Δ} is the average loss under the stress scenario. Eq. (5.12) can be applied on the entire portfolio or at subportfolio level. It is useful to note that the recovery process depends on several features, such as the nature of the credit and the country of its origin. Additionally, recovery flows are spread during the workout period and their estimate is subject to review until completion (with integral or partial recovery). Example 5.1 summarizes the relationship between specific provisions and recoveries.

Example 5.1 From a Default Flow to Specific Provisions and Cure

Bank Beta has a $100 million (performing) exposure at time t_0. At the same point in time, neither defaulted assets nor specific provisions are recorded. The probability of default under a stressed scenario is 2%, while the LGD is 40%. This LGD is estimated as an average over a population where 60% of credits are cured with no loss (i.e., LGD of 0%) and 40% have 100% loss. This corresponds to a 60% probability of cure and a 40% probability of noncure. Additionally, the recovery process is assumed to last 1 year (after default occurrence).

At the end of the first year (i.e., t_1), according to the above-mentioned parameters, a default flow of $2 million is estimated. At the same time, specific provisions of $0.80 million (i.e., LGD of 40%) are recorded in the balance sheet and a corresponding cost is noted in the profit and loss statement.

In line with the probability of cure described above, during the next time window (i.e., $[t_1, t_2]$) $1.20 million of cure credits flow back to the performing portfolio. In contrast, if no recovery occurs, the nonperforming exposure is nullified with the corresponding specific provisions (Fig. 5.2).

Example 5.1 From a Default Flow to Specific Provisions and Cure—cont'd

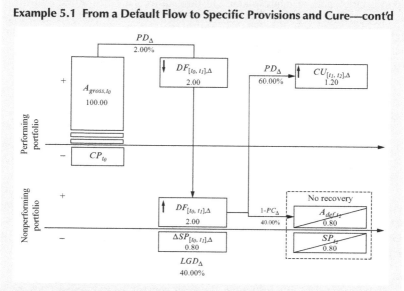

FIG. 5.2 Default flow, provisions, and cure over a 2-year time horizon.

In practice, a longer period may be necessary to complete the workout course described in Example 5.1. For this reason, some banks create non-performing credit subportfolios requiring a specific treatment. Exercise 5.1 outlines how to manage this process.

- **Write-off.** A write-off occurs when a bank realizes that an asset cannot be converted into cash or has no market value. It can be represented as follows:

$$WO_{i,two} = A_{i,t_{def}} + \sum_{t=1}^{two} RC_{i,t} - \sum_{t=1}^{two} RF_{i,t} - SP_{i,two}, \qquad (5.13)$$

where $A_{i,t_{def}}$ is the exposure at default and $RC_{i,t}$ stands for the recovery costs. Recovery flows $RF_{i,t}$ and specific provisions accumulated up to the moment of write-off ($SP_{i,two}$) are subtracted from the sum of $A_{i,t_{def}}$ and $\sum_{t=1}^{two} RC_{i,t}$.

Now that we have described the banking book projection (by focusing on the credit portfolio), the next step is to project the trading book, other assets, and liabilities.

5.2.4 Trading Book, Other Assets, and Liabilities Projection

The trading book of a commercial bank is usually functional to its treasury activity. As a consequence, balance sheet projections are linked to banking book dynamics. Nonetheless, one may consider a trading book to be driven by market opportunities and strategic plans. On this subject, a constrained optimization would be useful to model its evolution. A utility frontier should be dynamically defined according to risk appetite and macroeconomic conditions as shown below:

$$\begin{cases} TB_{t+h} = TB_t + \Delta TB_h, \\ \Delta TB_h = f(RA_t, \mathbf{x}_\Delta), \end{cases} \tag{5.14}$$

where TB_t represents the trading book volume at time t, and ΔTB_h is the variation of the trading book volume during the period h. This variation is linked to the risk appetite RA_t and macroeconomic conditions $\mathbf{x}_\Delta$. As anticipated, in commercial banking practice, a rolling assumption is commonly made. Variations are introduced as part of strategic planning.

Lending and fund raising need to be aligned in terms of overall magnitude and, to some extent, maturity. A bank manages asset and liability interactions to achieve sustainable economic growth and a liquidity balance. Interest rates are pivotal in lending and raising funds. The following components are imperative in projecting liabilities:

- liquidity mismatching between assets and liabilities;
- liability interest rate level and customer sensitivity to interest rates, reputation (rating), and so on;
- financial instruments through which to raise funds (e.g., deposits or bonds).

The following equation helps to summarize the key components of the liability projection process:

$$\begin{cases} L_{t+h} = L_t + \Delta L_h, \\ \Delta L_h = f(MSP_t, \mathbf{x}_\Delta), \end{cases} \tag{5.15}$$

where L_t represents the liability volume at time t, while ΔL_h is the variation of the volume during the period h. This variation is linked to market and strategic planning (MSP) and macroeconomic conditions $\mathbf{x}_\Delta$.

As per the trading book, a rolling assumption usually characterizes the liability projection. In this regard, one may refer to Chapter 3, where the balance between assets and liabilities was investigated from both the margin at risk and the liquidity perspectives.

Example 5.2 illustrates the assumptions and the strategic ingredients of an overall asset and liability growth process. It enables one to understand the core statements one needs to bear in mind when projecting a balance sheet.

Example 5.2 Asset and Liability Projection

A bank has total assets of $100 million at t_0. As part of the strategic plan to be applied during the 3-year period $[t_0, t_3]$, the following pivotal assumptions are considered:

Example 5.2 Asset and Liability Projection—cont'd

- **Risk appetite.** A reduction of traded securities is decisive in a strategy aimed at decreasing market risk. A preference for deposit against publicly issued bonds characterizes the liability risk profile.
- **Strategic planning.** A 3-year strategic plan embeds the sale of riskier corporate asset bonds to be partially replaced by government debt instruments. An increase in deposit volumes is pursued as an imperative goal to boost the bank's commercial share. This objective is pursued via geographical expansion accompanied by a moderate interest rate rise.
- **Macroeconomic environment (x_Δ).** A mild GDP reduction is assumed to affect the period $[t_0, t_3]$. Moreover, long-term interest rates are expected to be stable below the 2% threshold. Short-term rates are presumed to be stacked around 0.50%.

Table 5.1 outlines a stress testing asset and liability dynamic in line with the anchor points described above. The weight of the trading book is halved. Deposits increase by 5% and bonds reduce by 10% during the 3-year period under analysis.

TABLE 5.1 Asset and Liability Projection ($ Millions)

Assets		$[t_{-1}, t_0]$	$[t_0, t_1]$	$[t_1, t_2]$	$[t_2, t_3]$
	Securities at beginning of period		10.00	8.00	6.00
	Amortization		−3.00	−3.00	−3.00
	New business		1.00	1.00	2.00
	Securities at end of period	10.00	8.00	6.00	5.00
	Loans at beginning of period		90.00	93.00	96.00
	Amortization		−4.00	−5.00	−6.00
	New business		7.00	8.00	8.00
	Loans at end of period	90.00	93.00	96.00	98.00
	Total assets at end of period	100.00	101.00	102.00	103.00
Liabilities		$[t_{-1}, t_0]$	$[t_0, t_1]$	$[t_1, t_2]$	$[t_2, t_3]$
	Deposits at beginning of period		60.00	61.00	62.00
	Reductions		−4.00	−5.00	−5.00
	New business		5.00	6.00	6.00
	Deposits at end of period	60.00	61.00	62.00	63.00
	Bonds at beginning of period		30.00	29.00	28.00
	Amortization		−4.00	−5.00	−4.00
	New business		3.00	4.00	3.00
	Bonds at end of period	30.00	29.00	28.00	27.00
	Equity at beginning of period		10.00	11.00	12.00
	Net profit (loss)		1.00	1.00	1.00
	Equity at end of period	10.00	11.00	12.00	13.00
	Total liabilities at end of period	100.00	101.00	102.00	103.00

It is worth mentioning that in case of a static balance sheet assumption (EBA, 2016), a rolling scheme is followed for the trading book (as well as for other assets) and liabilities. Therefore the above assumptions are not part of the stress testing process.

In the next section, Bank Alpha's banking book, trading book, and liabilities are explored from a stress testing perspective.

5.2.5 Bank Alpha's Stress Testing Balance Sheet

The Bank Alpha example allows us to analyze the overall balance sheet projection process. As a first step of this process, the credit portfolio is examined. As detailed in Chapter 3, loans account for $70 billion net from provisions value at t_0 corresponding to $75 billion gross from provisions. On this latter, performing credits worth $62.40 billion and nonperforming credits worth $7.60 billion are recorded. Collective and specific provisions are detached from the gross exposure as detailed in Table 5.2.

The balance growth is estimated according to Eqs. (5.5) and (5.6), and the stress scenario described in Chapter 2 is used to make the projection. Credit risk parameters are shocked with the method described in Chapter 4. As a result, Table 5.3 summarizes the yearly projections along the entire stress testing horizon.

The balance growth is substantially stable throughout the stress exercise. Its amortization is higher during the first year $[t_0, t_1]$ than in following periods.

For the deterioration process, the first year $DF_{[t_0,t_1],\Delta}$ reaches $3.66 billion, compared with $3.03 billion and $2.99 billion for the following periods. This is

TABLE 5.2 Bank Alpha's Credit Portfolio at t_0 ($ Billions)

	A_{gross,t_0}	CP_{t_0}	A_{net,t_0}
Performing			
Corporate loans	22.00	−0.19	21.81
Retail loans	22.50	−0.24	22.26
Real estate loans	17.90	−0.17	17.73
Total performing	62.40	−0.60	61.80
	A_{gross,NP,t_0}	SP_{t_0}	A_{net,NP,t_0}
Nonperforming			
Corporate loans	6.00	−2.81	3.19
Retail loans	4.00	−1.26	2.74
Real estate loans	2.60	−0.33	2.27
Total nonperforming	12.60	−4.40	8.20
Total	75.00	−5.00	70.00

TABLE 5.3 Bank Alpha's Credit Portfolio Projection ($ Billions)

[t_0, t_1]

Performing

	A_{gross,t_0}	$BG_{[t_0,t_1],\Delta}$	$AM_{[t_0,t_1],\Delta}$	$DF_{[t_0,t_1],\Delta}$	$CU_{[t_0,t_1],\Delta}$	$A_{gross,t_1,\Delta}$	$CP_{t_1,\Delta}$	$A_{net,t_1,\Delta}$
Corporate loans	22.00	4.85	−1.12	−1.13	0.23	24.83	−0.25	24.58
Retail loans	22.50	1.98	−1.10	−1.14	0.24	22.48	−0.39	22.09
Real estate loans	17.90	0.98	−0.88	−1.40	0.30	16.90	−0.21	16.69
Total performing	62.40	7.81	−3.10	−3.67	0.77	64.21	−0.85	63.36

Nonperforming

	A_{gross,NP,t_0}	$WO_{[t_0,t_1],\Delta}$		$DF_{[t_0,t_1],\Delta}$	$CU_{[t_0,t_1],\Delta}$	$A_{gross,NP,t_1,\Delta}$	$SP_{t_1,\Delta}$	$A_{net,NP,t_1,\Delta}$
Corporate loans	6.00			1.13	−0.23	6.90	−3.03	3.87
Retail loans	4.00			1.14	−0.24	4.90	−2.53	2.37
Real estate loans	2.60			1.40	−0.30	3.70	−1.02	2.68
Total nonperforming	12.60			3.67	−0.77	15.50	−6.58	8.92
Total	75.00	7.81	−3.10			79.71	−7.43	72.28

[t_1, t_2]

Performing

	A_{gross,t_1}	$BG_{[t_1,t_2],\Delta}$	$AM_{[t_1,t_2],\Delta}$	$DF_{[t_1,t_2],\Delta}$	$CU_{[t_1,t_2],\Delta}$	$A_{gross,t_2,\Delta}$	$CP_{t_2,\Delta}$	$A_{net,t_2,\Delta}$
Corporate loans	24.83	3.94	−0.99	−1.04	0.34	27.08	−0.30	26.78
Retail loans	22.48	2.06	−0.90	−0.97	0.34	23.01	−0.43	22.58
Real estate loans	16.90	0.97	−0.68	−1.12	0.42	16.49	−0.23	16.26
Total performing	64.21	6.97	−2.57	−3.13	1.10	66.58	−0.96	65.62

(Continued)

TABLE 5.3 Bank Alpha's Credit Portfolio Projection ($ Billions)—cont'd

	A_{gross,NP,t_1}	$WO_{[t_1,t_2],\Delta}$		$DF_{[t_1,t_2],\Delta}$	$CU_{[t_1,t_2],\Delta}$	$A_{gross,NP,t_2,\Delta}$	$SP_{t_2,\Delta}$	$A_{net,NP,t_2,\Delta}$
Nonperforming								
Corporate loans	6.90			1.04	−0.34	7.61	−3.50	4.11
Retail loans	4.90			0.97	−0.34	5.53	−2.98	2.55
Real estate loans	3.70			1.12	−0.42	4.41	−1.26	3.15
Total nonperforming	15.50			3.13	−1.10	17.55	−7.74	9.81
Total	79.71	6.97	−2.57		–	84.13	−8.70	75.43

[t_2, t_3]

	A_{gross,t_2}	$BG_{[t_2,t_3],\Delta}$	$AM_{[t_2,t_3],\Delta}$	$DF_{[t_2,t_3],\Delta}$	$CU_{[t_2,t_3],\Delta}$	$A_{gross,t_3,\Delta}$	$CP_{t_3,\Delta}$	$A_{net,t_3,\Delta}$
Performing								
Corporate loans	27.08	3.45	−1.08	−1.10	0.31	28.66	−0.31	28.35
Retail loans	23.01	2.06	−0.92	−0.96	0.29	23.48	−0.42	23.06
Real estate loans	16.49	0.92	−0.65	−1.06	0.33	16.03	−0.20	15.83
Total performing	66.58	6.43	−2.65	−3.12	0.93	68.17	−0.93	67.24

	A_{gross,NP,t_2}	$WO_{[t_2,t_3],\Delta}$		$DF_{[t_2,t_3],\Delta}$	$CU_{[t_2,t_3],\Delta}$	$A_{gross,NP,t_3,\Delta}$	$SP_{t_3,\Delta}$	$A_{net,NP,t_3,\Delta}$
Nonperforming								
Corporate loans	7.61			1.10	−0.31	8.40	−3.77	4.63
Retail loans	5.53			0.96	−0.29	6.20	−3.26	2.94
Real estate loans	4.41			1.06	−0.33	5.14	−1.43	3.71
Total nonperforming	17.55			3.12	−0.93	19.74	−8.46	11.28
Total	84.13	6.43	−2.65		–	87.91	−9.39	78.52

mainly due to the steep shock characterizing the first year stress testing scenario described in Chapter 2. Write-offs are zero during the stress scenario. This follows from the assumption that specific provisioning expected values match the real ones. In this regard, misalignment between specific provisioning and actual losses usually occurs in practice. However, in this exercise, recoveries are assumed to align with credits net of provisions in a projection exercise based on averages.

Bank Alpha's entire balance sheet is shown in Table 5.4. Cash resources and the trading book do not significantly change during the stress testing exercise. An increase in the overall credit portfolio size characterizes the projection process. Deteriorated economic conditions together with expansive strategic plans cause a rise in loan exposure. On the liability side, the historical structure is replicated on the basis of stress testing projected assets. Low interest rates do not notably affect deposits. All other main items are assumed to maintain their magnitude at t_0.

As the final step of the projection, the next section illustrates the key assumptions underlying profit and loss evolution.

5.3 PROFIT AND LOSS PROJECTION

A close relation links the balance sheet to the profit and loss statement. Fig. 5.3 summarizes the key interconnections between these two statements within a stress testing mechanism. Firstly, a macroeconomic scenario is at the very top of the process. Then assets and liabilities are projected (as detailed in Section 5.2). As a following step, profits and losses are estimated. On this, the balance of positive and negative interest generates the NII. NIR and NIEs are added to the NII to compute the PPNR. The LICs account for the credit risk deterioration. Finally, other (minor) components and tax are considered to calculate the net profit (loss) as detailed in the lower part of the profit and loss box in Fig. 5.3.

A few additional elements deserve explanation. Firstly, an arrow drives net profit (loss) flows into equity. Secondly, operational losses feed the operational risk. Moreover, NII and NIR may be used to assess risk-weighted assets (RWAs) for operational risk when a bank relies on a standardized approach (i.e., arrow linking NII, NIR, and operational RWAs).

In what follows, profit and loss components are outlined, while Section 5.4 focuses on operational risk.

5.3.1 Profit and Loss Mechanics

Regulators only supply some general guidelines for the projection of profit and loss during a stress testing exercise. Therefore a bank is required to adopt its own method. This section summarizes good practice to conduct this estimate.

TABLE 5.4 Bank Alpha's Balance Sheet Projection for the Entire Stress Testing Period ($ Billions)

Assets		t_0		t_1		t_2		t_3
Cash resources			8.00		9.00		9.00		9.00
	Cash and due from banks	2.00		2.00		2.00		2.00	
	Interest-bearing deposits	6.00		7.00		7.00		7.00	
Securities			14.00		15.00		15.00		15.00
	Trading account	3.00		3.00		4.00		4.00	
	Available for sale	7.00		8.00		7.00		7.00	
	Held to maturity	4.00		4.00		4.00		4.00	
Loans			70.00		72.28		75.43		78.52
	Corporate loans	25.00		28.45		30.89		32.98	
	Retail loans	25.00		24.46		25.13		26.00	
	Real estate loans	20.00		19.37		19.41		19.54	
Other assets			8.00		9.02		10.19		11.66
Total assets			100.00		105.30		109.62		114.18

Liabilities		t_0		t_1		t_2		t_3
Deposits			70.00		75.04		78.31		81.40
	Noninterest bearing	5.00		5.04		5.31		8.40	
	Interest bearing	65.00		70.00		73.00		73.00	
Other liabilities			17.00		18.09		18.86		19.63
	Acceptances	3.00		3.00		3.00		3.00	
	Bonds	11.00		11.00		11.00		11.00	
	Other residual liabilities	3.00		4.09		4.86		5.63	
Subordinated debts			4.00		4.00		4.00		4.00
Noncontrolling interests			2.00		2.00		2.00		2.00
Shareholder equity			7.00		6.18		6.45		7.15
	Common shares	4.00		4.00		4.00		4.00	
	Preferred shares	1.00		1.00		1.00		1.00	
	Retained earnings	2.00		1.18		1.45		2.15	
Total liabilities			100.00		105.30		109.62		114.18

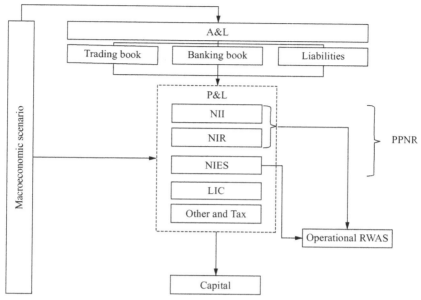

FIG. 5.3 Profit and loss stress testing framework. *A&L*, Assets and liabilities; *LICs*, loan impairment charge; *NIEs*, noninterest expenses; *NII*, net interest income; *NIR*, noninterest revenue; *P&L*, profit and loss; *PPNR*, preprovisioning net revenue; *RWAs*, risk-weighted assets.

The reader is invited to refer to the previous chapters for a comprehensive understanding of the mechanism. In this regard, Fig. 5.3 may be used as a reference for the itinerary based on the following elements: NII, NIR, NIEs, and LICs. In line with the previous sections, h is assumed to be 1 year.

- **NII ($NII_{h,\Delta}$).** As detailed in Chapter 3, the NII can be represented as follows:

$$
\begin{aligned}
NII_{h,\Delta,BG} = \sum_{t=0}^{h} \Bigg(& \sum_{i=1}^{N} \mathbb{1}_{i,rep,t} A_{i,ra,t} C_{i,\Delta,t} DD_{i,t} \\
& + \sum_{i_{BG}=1}^{N_{BG}} \mathbb{1}_{i_{BG},rep,t} A_{i_{BG},ra,t} C_{i_{BG},\Delta,t} DD_{i_{BG},t} \Bigg) \\
- \sum_{t=0}^{h} \Bigg(& \sum_{j=1}^{M} \mathbb{1}_{j,rep,t} L_{j,ra,t} C_{j,\Delta,t} DD_{j,t} \\
& + \sum_{j_{BG}=1}^{M_{BG}} \mathbb{1}_{j_{BG},rep,t} L_{j_{BG},ra,t} C_{j_{BG},\Delta,t} DD_{j_{BG},t} \Bigg),
\end{aligned}
\tag{5.16}
$$

where $NII_{h,\Delta,BG}$ includes the balance growth described in the previous sections. Interest coupons $C_{i,\Delta,t}$, $C_{iBG,\Delta,t}$, $C_{j,\Delta,t}$, and $C_{jBG,\Delta,t}$ are estimated according to a stressed term structure of interest rates. Given the multiperiod nature of a stress test, Eq. (5.16) holds for each subperiod.

- **NIR** ($NIR_{h,\Delta}$). A series of revenue streams are collected under the NIR umbrella. Fees and commission related to debt capital markets, trade services, deposits, debit and credit cards, and similar instruments are in the scope of NIR. An explicit link between NIR and macroeconomic variables needs to be assessed to perform a coherent stress testing exercise. On this subject, a business segmentation is often required to cluster NIRs according to their relevance (e.g., wholesale, retail, and capital market fees, commission, and expenses). The following regression equation may be applied to each (rk) revenue stream:

$$NIR_{rk,t} = \beta_{rk,0} + \beta_{rk,1}x_{1,t} + \cdots + \beta_{rk,p}x_{p,t} + \epsilon_{rk,t}, \qquad (5.17)$$

where, as usual, $\mathbf{x}_t$ is the vector of macroeconomic variables, while errors are i.i.d. normally distributed. It is worth noting that, as detailed for balance projections, more sophisticated statistical techniques may be used. Moreover, overlays are applied in practice to align the stress testing evolution with strategic plans (under a given macroeconomic scenario). NIR may be represented as follows over a 1-year stress testing interval:

$$NIR_{h,\Delta} = \sum_{rk=1}^{RK} NIR_{rk,h}(\mathbf{x}_\Delta), \qquad (5.18)$$

where $NIR_{rk,h}(\mathbf{x}_\Delta)$ serves as an NIR stream over the period h.

- **NIEs** ($NIE_{h,\Delta}$). NIEs include costs that are instrumental to banking business. These expenses are usually stable over a multiyear period. Large changes may occur when a bank decides to cease a business, exit a country, and so on. An all-embracing representation of NIEs highlights the combination of different operating costs as given below:

$$NIE_{h,\Delta} = \sum_{ek=1}^{EK} NIE_{ek,h}(\mathbf{x}_\Delta), \qquad (5.19)$$

where $NIE_{ek,h}(\mathbf{x}_\Delta)$ represents the operating costs over the interval h, given macroeconomic stressed conditions. One needs to bear in mind that NIEs are usually very stable. Thus it is common practice not to develop any statistical model but to use the usual strategic planning approach to make projections. The sum of NII, NIR, and NIEs corresponds to the PPNR.

- **LICs.** Impairment charges represent the movement in provisions (collective and specific) plus any additional write-off occurring at the end of the recovery process. The main components of LICs are as follows:

- **Collective provisions' flow** ($\Delta CP_{h,\Delta}$). The collective provisions' flow (related to the performing portfolio) is estimated as follows:

$$\Delta CP_{h,\Delta} = CP_{t+h} - CP_t. \qquad (5.20)$$

- **Specific provisions' flow** ($\Delta SP_{h,\Delta}$). The nonperforming portfolio provisions' flow is estimated as follows:

$$\Delta SP_{h,\Delta} = SP_{t+h} - SP_t. \qquad (5.21)$$

- **Write-off** ($WO_{h,\Delta}$). As detailed in Eq. (5.13), the write-off is related to the excess cost (in a broad sense) due to nonrecovered credits.

  To conclude, LICs for the entire portfolio can be represented as follows:

$$LIC_{h,\Delta} = \Delta CP_{h,\Delta} + \Delta SP_{h,\Delta} + WO_{h,\Delta}. \qquad (5.22)$$

Exercise 5.2 shows how to represent the credit life cycle deterioration from an accounting perspective.

The next section focuses on Bank Alpha to summarize the profit and loss stress testing mechanism.

5.3.2 Bank Alpha's Stress Testing Profit and Loss

In what follows, Bank Alpha's key profit and loss components are investigated. In line with the previous section, a distinction is made among each of the key profit and loss elements.

- **NII** ($NII_{h,\Delta}$). A reduction of interest rates characterizes the stress testing scenario described in Chapter 2. The margin at risk analysis performed in Chapter 3 highlighted an NII reduction of $0.10 million over the period $[t_0, t_1]$. The change in mix due to the planned balance sheet evolution causes Bank Alpha to suffer a slightly bigger NII reduction over the same interval. Indeed, starting from the nonstressed $2.50 billion 1-year NII, Table 5.5 shows a $2.31 billion NII over the same period. This corresponds to a $0.19 billion reduction. NII slightly increases to $2.32 billion over the following 1-year intervals $[t_1, t_2]$ and $[t_2, t_3]$.
- **NIR** ($NIR_{h,\Delta}$). Table 5.6 shows a stable path of NIR of around $2.30 billion during the 3-year stress testing exercise. Fees and commission moderately reduce from the $2.45 billion estimate under a nonstressed scenario.
- **NIEs** ($NIE_{h,\Delta}$). NIEs are almost insensitive to macroeconomic changes. More specifically, Table 5.7 highlights that Bank Alpha's NIEs are stable at around $3 billion during the stress period (i.e., the same level as for no stress).
- **PPNR** ($PPNR_{h,\Delta}$). Bank Alpha's PPNR is not severely affected by the macroeconomic scenario. This is mainly due to its balanced asset and liability structure. Severe changes would have affected a disproportioned

TABLE 5.5 Stress Testing of Bank Alpha's Net Interest Income ($ Billions)

		Asset interest					
		[t_0, t_1]		[t_1, t_2]		[t_2, t_3]	
Cash resources			0.16		0.18		0.18
	Cash						
	Interest-bearing deposits	0.16		0.18		0.18	
Securities			0.41		0.42		0.40
	Trading account	0.07		0.06		0.08	
	Available for sale	0.21		0.23		0.20	
	Held to maturity	0.13		0.13		0.12	
Loans			3.28		3.20		3.27
	Corporate loans	1.36		1.38		1.44	
	Retail loans	1.09		1.02		1.03	
	Real estate loans	0.83		0.80		0.80	
Other assets							
Total assets			3.85		3.80		3.85
Average interest rate			3.85%		3.67%		3.61%

		Liability interest					
		[t_0, t_1]		[t_1, t_2]		[t_2, t_3]	
Deposits			−1.25		−1.24		−1.29
	Noninterest bearing						
	Interest bearing	−1.25		−1.24		−1.29	
Other liabilities			−0.23		−0.19		−0.19
	Acceptances	−0.02		−0.02		−0.02	
	Bonds	−0.19		−0.16		−0.16	
	Other residual liabilities	−0.01		−0.01		−0.02	
Subordinated debts			−0.06		−0.05		−0.05
Noncontrolling interests							
Total liabilities			−1.54		−1.48		−1.53
Average interest rate			−1.72%		−1.54%		−1.54%
Net interest income			**2.31**		**2.32**		**2.32**

TABLE 5.6 Stress Testing of Bank Alpha's Noninterest Revenue ($ Billions)

Noninterest Revenue			
	$[t_0, t_1]$	$[t_1, t_2]$	$[t_2, t_3]$
Fee and commission income	2.45	2.46	2.47
Fee and commission expenses	−0.16	−0.16	−0.16
Noninterest revenue	2.29	2.30	2.31

TABLE 5.7 Stress Testing of Bank Alpha's Noninterest Expenses ($ Billions)

Noninterest Expenses			
	$[t_0, t_1]$	$[t_1, t_2]$	$[t_2, t_3]$
Staff expenses	−1.38	−1.37	−1.39
IT and infrastructure	−1.31	−1.31	−1.31
Other expenses	−0.30	−0.30	−0.29
Noninterest expenses	−2.99	−2.98	−2.99

TABLE 5.8 Stress Testing of Bank Alpha's Preprovisioning Net Revenue ($ Billions)

Preprovisioning Net Revenue			
	$[t_0, t_1]$	$[t_1, t_2]$	$[t_2, t_3]$
Net interest income	2.31	2.32	2.32
Noninterest revenue	2.29	2.30	2.31
Noninterest expenses	−2.99	−2.98	−2.99
Preprovisioning net revenue	1.61	1.64	1.64

bank by influencing NII and NIR. Table 5.8 shows a stable PPNR of around $1.6 billion compared with $2 billion expected in the absence of stress.

- **LICs** ($LIC_{h,\Delta}$). Credit deterioration is one of the key issues in a stress test. This is particularly true in commercial banking, where business mainly relates to credit quality. Table 5.9 provides a summary of Bank Alpha's LIC evolution during the stress testing exercise. Its overall LICs in absence of stress are $0.55 billion. The reader is invited to refer to Chapter 3, where a

TABLE 5.9 Stress Testing of Bank Alpha's Loan Impairment Charges ($ Billions)

Loan Impairment Charges			
	$[t_0, t_1]$	$[t_1, t_2]$	$[t_2, t_3]$
Collective provisioning	−0.25	−0.11	0.03
Specific provisioning	−2.18	−1.16	−0.72
Write-off	–	–	–
Loan impairment charges	−2.43	−1.27	−0.69

TABLE 5.10 Bank Alpha's Profit and Loss Evolution for the Entire Stress Testing Period ($ Billions)

Profit and Loss			
	$[t_0, t_1]$	$[t_1, t_2]$	$[t_2, t_3]$
Net interest income	2.31	2.32	2.32
Noninterest revenue	2.29	2.30	2.31
Noninterest expenses	−2.99	−2.98	−2.99
Preprovisioning net revenue	1.61	1.64	1.64
Loan impairment charges	−2.43	−1.27	−0.69
Tax	–	−0.10	−0.25
Net profit/loss	−0.82	0.27	0.70

nonstress profit and loss is summarized (i.e., Table 3.2). Table 5.9 highlights a significant impact on both collective provisions and specific provisions. The default flow over the first period $[t_0, t_1]$ captures almost immediately the stress peak. During the following periods (i.e., $[t_1, t_2]$ and $[t_2, t_3]$) the crisis is partially reabsorbed and credits are affected by a renewed economic confidence.

- **Net profit/loss.** Table 5.10 outlines Bank Alpha's profit and loss evolution during the stress test. Chapter 3 showed a nonstressed $1.06 billion profit. Adverse macroeconomic conditions reduce Bank Alpha's capability to generate profits, causing a −$0.82 billion result for the period $[t_0, t_1]$. The following two periods record positive results. Indeed, $[t_1, t_2]$ and $[t_2, t_3]$ account for profits of $0.27 billion and $0.70 billion, respectively. These results feed the retained earnings row of Table 5.4.

The next section introduces the core operational risk ideas. The role of historical and projected losses is highlighted in connection with profit and loss stress testing. RWAs are also outlined as a key element of pillar 1 regulatory capital.

5.4 CONDUCT AND OPERATIONAL RISK STRESS TESTING

Conduct risk may be defined as the current or prospective risk of losses arising from an inappropriate supply of financial services including cases of willful or negligent misconduct. Basel II (BIS, 2006) defines operational risk as the risk of loss resulting from inadequate or failed internal processes, people, and systems or from external events. This definition includes legal risk but excludes strategic and reputational risks. Legal risk includes, but is not limited to, exposure to fines, penalties, or punitive damages resulting from supervisory actions, as well as private settlements.

Banks are required to project the profit and loss impact of losses arising from conduct risk and other operational risks. Projections should take into account the economic and financial environment, when relevant. The same applies to RWAs as detailed below.

5.4.1 Projection of Conduct and Operational Losses

Qualitative or quantitative approaches may be used to project conduct and operational risk losses. Banks estimate future costs on the basis of historical material events and new potential risks related to internal and external environment.

A range of outcomes is assessed, and for each event a probability is assigned. Then an overall assessment takes into account the probability of each event as well as its economic impact. Conduct and operational losses are expected to exceed their corresponding provisions during a stress testing exercise. This occurs except for events where there is a high degree of certainty over the estimated cost.

Table 5.11 summarizes the Basel II operational risk event type. Conduct risk may have some overlap with some of these categories, in particular with internal and external fraud.

All in all, a specific assessment is usually required to project these risks. In all cases, a reconciliation is imperative with the NIE estimates described in the previous section. In what follows, the focus is on RWAs for operational risk to be analyzed as core component of the pillar 1 capital requirement.

5.4.2 Risk-Weighted Assets for Operational Risk

There are two main ways to assess operational risk. The first relies on high-level balance sheet measures. The second is event based. With regard to the balance sheet approach, operational losses are quantified on a macro level. No attempt is made to identify the events or causes of losses. In contrast, event-based models quantify operational risks on a micro level and are based on the identification of internal events. The advantage of the latter approach lies in

TABLE 5.11 Basel Operational Risk Event Type Summary

	Basel II Event Type
1. Internal fraud	Misappropriation of assets, tax evasion, intentional mismarking of positions, bribery
2. External fraud	Theft of information, hacking damage
3. Employment practices and workplace safety	Discrimination, worker compensation, employee health and safety
4. Clients, products, and business practice	Market manipulation, antitrust, improper trade, product defects, fiduciary breaches, account churning
5. Damage to physical assets	Natural disasters, terrorism, vandalism
6. Business disruption and system failures	Utility disruptions, software failures, hardware failures
7. Execution, delivery, and process management	Data entry errors, accounting errors, failed mandatory reporting, negligent loss of client assets

a profound understanding of operational risk events (i.e., why and how these events occurred).

Basel II distinguishes between the basic indicator approach (BIA), the standardized approach (SA), and the advanced measurement approach (AMA) as described below:

- **Basic indicator approach (BIA).** Under the BIA, the simplest approach relies on the gross income as a sum of NII and NIR. It serves as a proxy for the scale of operational risk of the bank. Hence the bank must hold capital to cover potential operational risks equal to the average over the previous 3 years of a fixed percentage (e.g., 15%) of positive annual gross income. The operational risk RWA can be described as follows

$$\text{RWA}_{or,\text{BIA}} = 12.5 \frac{15\% \sum_{t=1}^{3} \mathbb{1}_t GI_t}{\sum_{t=1}^{3} \mathbb{1}_t}, \quad (5.23)$$

where the subscript *or* stands for operational risk, while *GI* is the gross income referred to previous years and

$$\mathbb{1}_t = \begin{cases} 1 & \text{if } GI_t \geq 0 \\ 0 & \text{otherwise.} \end{cases}$$

- **Standardized approach (SA).** With regard to the SA, the activities of a bank are divided into eight distinct business lines. For each of them the gross income is considered as a broad indicator of the operational risk exposure.

As detailed in Table 5.12, the capital requirement weight ranges from 12% to 18% (denoted as β_{bl}) of the business line gross income as detailed below:

$$\text{RWA}_{\text{or,SA}} = 12.5 \frac{\sum_{t=1}^{3} \max(\sum_{bl=1}^{8} GI_{bl,t}\beta_{bl}, 0)}{3}, \tag{5.24}$$

where *SA* refers to the standardized approach. $GI_{bl,t}$ is the gross income of year *t* for the business line. β_{bl} is a fixed percentage of $GI_{bl,t}$ for each of eight business lines.

TABLE 5.12 Standardized Approach Operational Risk Weights

	β_{bl} (%)
Corporate finance	18
Trading and sales	18
Retail banking	12
Commercial banking	15
Payment and settlement	18
Agency services	15
Asset management	12
Retail brokerage	12

- **Advanced measurement approach (AMA).** Under the AMA, the capital requirement is equal to the risk measure generated by the bank's internal operational risk measurement system. The bank is required to meet certain qualitative and quantitative standards to qualify for use of the AMA. As per the use of market and credit internal risk models to assess the capital requirement, the use of the AMA is subject to supervisory approval. Most banks use a combination of two AMAs to measure operational risk. A distinction holds between the loss distribution approach and the scorecard approach. The first relies on quantitative statistical methods to analyze historical loss data. The second relies mainly on qualitative measures to assess potential operational losses. These approaches complement each other. As a historical data analysis is backward looking and quantitative, the scorecard approach encompasses forward-looking and qualitative indicators.

The idea of unexpected loss introduced in Chapter 4 is used also for operational risk as detailed in Fig. 5.4.

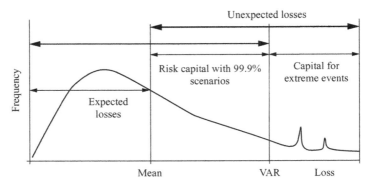

FIG. 5.4 Operational risk loss distribution under the advanced measurement approach. *VAR*, Value at risk.

The capital should cover both expected losses (e.g., in the form of provisions) and unexpected losses. With regard to the events generating operational risks, Basel II proposes the classification described in Table 5.11. These event categories are used to develop quantitative and qualitative models.

The aggregate distribution for each event type is obtained as the convolution of the frequency and the severity of loss distributions. An assumption is required to specify the frequency of loss arising from each event type (e.g., Poisson distribution), while another distribution (e.g., lognormal) is used to model the severity of losses. The operational risk advanced modeling can be described as follows:

- **Loss distribution.** The aggregated distribution of losses is estimated via Monte Carlo simulations for each event type.
- **Total risk.** The overall risk is computed by use of a given operational risk dependence structure. In what follows, two extreme alternatives are considered in addition to one based on real data estimates:
  * **Independence among event types.** The total capital charge is obtained by the summing up of risk charges for each event type.
  * **Realistic dependence.** Real-world dependencies are usually estimated via copulas.

The next section outlines Bank Alpha's operational risk during the stress testing exercise.

5.4.3 Bank Alpha's Stress Testing Operational RWA

Bank Alpha uses the standardized approach. On this subject, the weights in Table 5.12 are applied on the gross income of each business line. Table 5.13 summarizes Bank Alpha's operational RWAs based on the average margins

TABLE 5.13 Bank Alpha's Risk-Weighted Assets for Operational Risk ($ Billions)

	RWAs t_0	RWAs t_1	RWAs t_2	RWAs t_3
Corporate finance	0.11	0.11	0.11	0.10
Trading and sale	0.56	0.55	0.53	0.52
Retail banking	2.40	2.34	2.28	2.22
Commercial banking	4.22	4.11	4.00	3.90
Payment and settlement	0.90	0.88	0.85	0.83
Agency services	0.28	0.27	0.27	0.26
Asset management	0.23	0.22	0.21	0.21
Retail brokerage	0.23	0.22	0.21	0.21
Total	8.93	8.69	8.46	8.24

RWAs, Risk-weighted assets.

computed for the 3-year period before the initial date of the stress testing. The profit and loss described in the previous section is used to measure the gross income along the stress testing horizon.

5.5 SUMMARY

Balance growth is a crucial element in a dynamic planning as well as in stress testing. The credit portfolio is one of the key areas regulators are keen to investigate. For this reason, an extensive study of the credit life cycle characterized the first part of the chapter. On this subject, a distinction was highlighted between static and dynamic balance sheet approaches. A multiperiod framework to project assets and liabilities was introduced by our following the dynamic approach. The macroeconomic scenario described in Chapter 2 was used to project Bank Alpha's balance sheet and estimate its profit and loss dynamic under stress. The impacts of adverse scenarios were examined by our paying particular attention to PPNR and LICs. Finally, a brief analysis of conduct and operational losses showed their implications in a stress testing exercise.

SUGGESTIONS FOR FURTHER READING

This chapter covered a hybrid area by connecting strategic planning, risk management, and regulatory requirements. A useful stress testing methodological guide is provided in EBA (2016), where all topics examined so far are discussed in terms of high-level prescriptions. Following a similar approach, the FRB (2015) and the BOE (2015) describe methods to be applied in a dynamic balance sheet setting.

Any practitioner interested in operational risk may find Soprano et al. (2009), Girling (2013), and Blunden and Thirlwell (2013) useful.

EXERCISES

Exercise 5.1 A bank has an exposure of \$200 million. At time t, neither defaulted assets nor specific provisions are recorded. The probability of default under a stressed scenario is 3%. The LGD for new flows of default is 28% (LGD_{Δ}^{NEW}). The workout process has a 3-year length, and credits are cured within the first year. The probability of cure (PC) is 30% (consequently 70% is the probability of noncure). At the same time, a nonnull recovery characterizes the noncured credit stock. More precisely, an LGD of 40% needs to be applied to the old stock of defaulted credits (i.e., $LGD_{\Delta}^{OLD} = 40\%$). Describe the work flow and assess what follows.

- Default flow: $DF_{[t_0,t_1],\Delta}$.
- Variation of specific provisions stock: $\Delta SP_{[t_0,t_1]}$.
- Cure rate flow: $CU_{[t_1,t_2]}$.
- Defaulted asset stock in t_2: A_{def,t_2}.
- Specific provisions stock in t_2: SP_{t_2}.
- Recovery flows: RF.

TABLE 5.14 Business Line's Gross Income ($ Millions)

	t_{-2}	t_{-1}	t_0
Corporate finance	10.00	10.00	10.00
Trading and sale	20.00	−60.00	30.00
Retail banking	20.00	20.00	30.00
Commercial banking	20.00	15.00	10.00
Payment and settlement	10.00	−40.00	10.00
Agency services	20.00	15.00	
Asset management		20.00	30.00
Retail brokerage	−10.00	10.00	20.00

Exercise 5.2 (LICs). On the basis of the specific provisioning process described in Exercise 5.1, represent the balance sheet and profit and loss dynamic along buckets $t_0, \ldots, t_5$ by use of the common T-form accounting representation.

Exercise 5.3 Consider the gross income of a bank described in Table 5.14 during years t_1, t_2, and t_3. Compute the corresponding standardized RWAs.

REFERENCES

BIS, 2006. Basel II International Convergence of Capital Measurement and Capital Standards: A Revised Framework, Bank for International Settlements, Basel.

Blunden, T., Thirlwell, J., 2013. Operational Risk, second ed. Pearson, Harlow.

BOE, 2015. Stress testing the UK banking system: key elements of the 2015 stress test, Bank of England Publications, London.

EBA, 2016. 2016 EU-wide stress test.

FRB, 2015. Comprehensive Capital Analysis and Review 2015: assessment framework and results, Board of Governors of the Federal Reserve System, Washington, DC.

Girling, P., 2013. Operational Risk Management, Wiley, Hoboken.

Krzanowski, W.J., 2000. Principles of Multivariate Analysis, Oxford University Press, Oxford.

Soprano, A., Crielaard, B., Piacenza, F., Ruspantini, D., 2009. Measuring Operational and Reputational Risk, Wiley, Chichester.

Chapter 6

Regulatory Capital, RWA, Leverage, and Liquidity Requirements Under Stress

Chapter Outline

Growing attention has been devoted in recent years to banks' capital adequacy, liquidity willingness, and balance sheet integrity.

Banks are required both to improve the quality of their resources and to enforce their capital ratios. This chapter brings together all components examined in the previous chapters to compute synthetic measures of solvency. Additionally, it builds a bridge toward the risk integration and reverse stress testing processes described in Chapters 7 and 8.

Regulators use capital ratios to assess bank resilience. In this regard, it is useful to investigate how regulatory capital is defined within the Basel III Accord frame. Starting from an accounting representation, we apply a series of adjustments to compute regulatory capital layers. The differentiation among these tiers is based on the degree of dilution of own funds compared with equity (i.e., common shares, retained earnings, and so on).

Stress Testing and Risk Integration in Banks. http://dx.doi.org/10.1016/B978-0-12-803590-0.00006-0

Market, credit, and operational RWAs are studied starting from a silo standpoint. Stressed risk parameters, balance sheet projections, and profit and loss estimates are critical gears of this engine. On this subject, after a silo RWA computation, an additional step is required to align with regulatory constraints on an aggregated basis. All in all, the analysis of capital ratios finalizes the end-to-end stress testing process examined through the illustrative example of Bank Alpha.

The recent financial crisis led regulators to introduce nonrisk-weighted measures to assess a bank's capability to face adverse conditions. The last part of the chapter shows how to monitor the balance between resources and investments by means of the leverage ratio. Furthermore, liquidity is studied through the liquidity coverage ratio (LCR) and the net stable funding ratio (NSFR).

6.1 INTRODUCTION

One may be dragged into the details of a specific topic within a wide stress testing exercise and miss the end-to-end picture behind it. The goal of this chapter is to wrap up the knowledge gained in the previous chapters and present stress testing core results.

Chapters 2–5 outlined a series of methods, processes, and instruments used within a stress testing exercise. Market, credit, and operational risks polarized the attention. Nonetheless, a wider perspective was followed by our considering a broader range of risks beyond pillar 1 (e.g., interest rate and liquidity). It is now time to show how to use these tools to assess the overall bank resilience against adverse conditions. A regulatory perspective characterizes this chapter. In contrast, Chapters 7 and 8 will relax these restrictions by applying a wider managerial view.

One of the main ideas underlying the Basel II Accord is to ensure banks have enough capital to face unexpected losses. For this purpose, the following threshold was introduced as a trigger for capital enforcement:

$$\text{Regulatory capital} \geq 8\%(RWA_{market} + RWA_{credit} + RWA_{operational}), \quad (6.1)$$

where $RWA_{market} + RWA_{credit} + RWA_{operational}$ is the sum of market, credit, and operational RWAs.

One should question why we should use RWAs instead of assets (as is common practice in finance) in Eq. (6.1). The main reason is to take into account the impact of risks. In other words, assuming that a risky asset has a 100% weight, a \$100 investment on such an asset involves a minimum regulatory capital of \$8. In contrast, a risk-free asset having a 0% weight does not need any capital exceeding the expected losses requirement (already included within own funds by means of provisions).

Fig. 6.1 helps us understand the relationship between assets, liabilities, and regulatory capital. Assets are classified according to their risk profile to compute the RWAs. The latter may be less than or greater than the total asset value (for

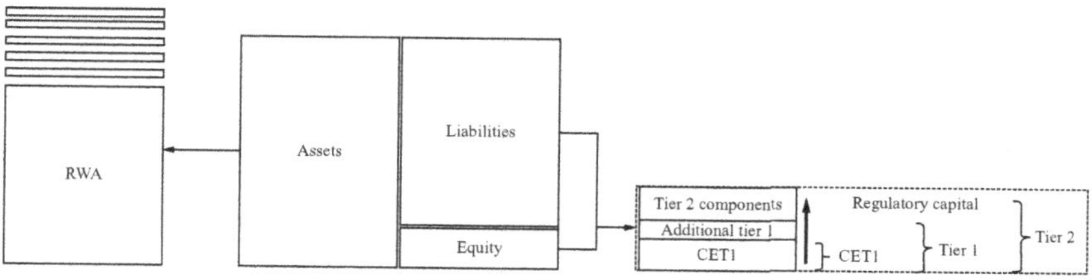

FIG. 6.1 Regulatory framework at a glance. *RWAs*, Risk-weighted assets.

this reason the RWA box has some thin layers on top). In contrast, regulatory capital components are ranked in line with their capability to absorb losses. Therefore common equity tier 1 (CET1) is made up of instruments such as common shares and retained earnings. Debt-like instruments with some degree of subordination are included as additional tier 1 and tier 2 components.

In line with this representation, a risk-based capital ratio is computed as follows

$$\text{Capital ratio} = \frac{\text{Regulatory capital}}{\text{RWAs}}, \qquad (6.2)$$

where higher ratios ensure a more comprehensive risk coverage. The connection between Eqs. (6.1) and (6.2) is evident. The 8% rule applies to a total capital ratio definition as described in Section 6.3.2.

The crisis experienced in 2007–09 led a regulatory enhancement. The vast bulk of Basel II remaining in operation, changes due to Basel III (BIS, 2011) focused on few specific areas. Firstly, the credit risk was explicitly included within the market risk area (e.g., counterpart credit risk, credit value adjustments). Then improvements on capital standards as well as the introduction of leverage and liquidity ratios completed the renewed frame.

As part of the scheme described by Eqs. (6.1) and (6.2), Basel III revises the definition of capital and specifies new minimum capital requirements in line with Table 6.1. By Jan. 2019, the total minimum total capital requirement will increase from 8% to 10.50%. This is due to a 2.50% capital conservation buffer to absorb losses during periods of financial and economic stress. Additional buffers may be required for stress testing purposes as described in Chapter 1. As a bank's capital falls into the buffer range and approaches the minimum requirement, the bank would be subject to increasing restrictions on earnings distribution. Basel III also establishes a countercyclical capital buffer and another cushion needs to be considered for systemically important financial institutions. Currently there is a lot of uncertainty regarding what will be classified as a systemically important financial institution and what the consequences will be. The countercyclical capital buffer is structured as an add-on to the capital conservation buffer. It is meant to counterbalance procyclical bank lending behavior. This is achieved by the linking of the height of this buffer to the economic cycle. If there are signs of excessive credit growth, the buffer

TABLE 6.1 Basel Committee on Banking Supervision Road Map for Minimum Capital Requirements (the Dates Refer to Jan. 1)

	2016	2017	2018	2019
Minimum common equity capital ratio (%)	4.50	4.50	4.50	4.50
Capital conservation buffer (%)	0.625	1.25	1.875	2.50
Minimum common equity plus capital conservation buffer (%)	5.125	5.75	6.375	7.00
Minimum tier 1 Capital (%)	6.00	6.00	6.00	6.00
Minimum total capital (%)	8.00	8.00	8.00	8.00
Minimum total capital plus conservation buffer (%)	8.625	9.25	9.875	10.50

can be applied at the discretion of the national regulatory authority. The buffer ranges from 0% to 2.5% and consists of either tier 1 common equity or other fully loss absorbing capital instruments.

Section 6.2 focuses on the numerator of Eq. (6.2), while Section 6.3 highlights how to compute the denominator. In this regard, market, credit, and operational risk silos are aggregated to compute the overall RWAs.

Furthermore, as a response to 2007–09 crisis, Basel III introduced nonrisk-based indicators to monitor bank solvency. On this subject, Section 6.4 describes the leverage ratio as an indicator of a structural balance between own resources and assets. Following a liquidity view, a distinction is made between coverage and the NSFR: the former attempts to capture the short-term mismatch between cash outflows and inflows, whereas the latter focuses on a longer-term liquidity counterbalance.

Unlike in the previous chapters, a process-oriented approach qualifies the following sections. The focus is on regulatory rules to compute the synthetic indices of endurance described above. In this regard, links between regulatory requirements and accounting rules are crucial for a comprehensive understanding of the entire method.

6.2 REGULATORY CAPITAL

Basel I (BIS, 1988) linked the minimum capital standards to credit risk. An amendment is extended to market risk the capital requirement computation. Then, Basel II (BIS, 2006) intended to enhance international capital standards by means of internal ratings-based (IRB) methods to assess the credit risk. Operational risk was also included as an additional component of the pillar 1 risk framework. In line with the above, Basel III strengthened the rules on capital (and introduced nonrisk-based indices based on structural leverage and liquidity). The following sections describe how to compute the regulatory capital starting from a financial reporting (accounting) representation. A stress testing perspective is followed and illustrated through Bank Alpha.

6.2.1 How to Compute the Regulatory Capital

Regulatory capital rules do not necessarily align with accounting standards. Hence an adjustment is required to extrapolate regulatory capital figures from accounting reports. As a starting point, one may think of Bank Alpha's equity introduced in Chapter 3. In this case, shareholder equity is made up of common shares, preferred shares, and retained earnings. Some additional components may be considered as part of the regulatory capital by our distinguishing among CET1, additional tier 1 components, and tier 2 capital as listed below:

- **CET1.** Common equity tier 1 consists of a combination of shares and retained earnings. It is the primary and most restrictive form of regulatory capital. To qualify as tier 1, capital has to be subordinated, perpetual in nature, loss bearing, and fully paid up with no funding having come from the bank. On top of common equity, all major adjustments are applied as described below.
- **Additional tier 1 capital.** This additional layer of the tier 1 capital consists of instruments paying discretionary dividends having neither a maturity date nor an incentive to redeem them. Innovative hybrid capital instruments are phased out because of their fixed distribution percentage, nonloss absorption capabilities, and incentive to redeem them through features such as step-up clauses.
- **Tier 2 capital.** Tier 2 capital contains instruments that are capable of bearing a loss, not only in the case of default but also in the event that a bank is unable to support itself in the private market. Their contractual structure needs to allow banks to write them down or convert them into shares.

Table 6.2 outlines some of the differences between Basel II and Basel III components. This is an illustrative and nonexhaustive exemplification.

Misalignments may arise between accounting and regulatory capital requirements. On this subject, one needs to apply regulatory adjustments. As an example, goodwill and intangibles are usually recognized under the current accounting standards. In contrast, these elements need to be excluded from the regulatory capital. Likewise, the (negative) difference between accounting provisions and the IRB expected loss (i.e., estimated on internal rating parameters) is deducted from the regulatory capital. Table 6.3 shows some illustrative examples of adjustments.

A two-step process applies to compute capital requirements. Firstly, accounting components are qualified in terms of regulatory categories (i.e., CET1, additional tier 1 capital, and tier 2 capital). Secondly, deductions are

TABLE 6.2 Comparison Between Basel II and Basel III Regulatory Capital Treatment of Some (Major) Capital Components

Capital Category	Basel II	Basel III
Common shares	CET1	CET1
Retained earnings	CET1	CET1
Innovative capital instruments	Additional tier 1 capital	Excluded and grandfathered
Noninnovative capital instruments	Included in additional tier 1 capital, subject to conditions	
Subordinated debt	Tier 2 capital	Included in tier 2 capital, subject to conditions

TABLE 6.3 Regulatory Capital Adjustments (Illustrative Examples)

Adjustment	Description
Goodwill and intangible assets	To be deducted from CET1
Shortfall provisions	To be deducted from CET1
Unrealized gains and losses	To be taken into account in CET1

computed. A fully comprehensive description of all cases one may encounter in practice goes beyond the scope of this section. The two phases are outlined by our focusing on the key elements of Bank Alpha's illustrative example as listed below:

1. **Identification of regulatory categories.** As part of the first step, balance sheet components are examined and qualified as tier 1 or tier 2 as follows:

   • **Shareholder equity:**
     – **Common shares.** Common stockholder equity is included in accounting capital and consists of voting shares. This is the most desirable capital element from a supervisory perspective. Indeed, shares absorb bank losses commensurate with their accounting value. They provide a savings association with the maximum amount of financial flexibility necessary during a crisis and are qualified as CET1.
     – **Preferred shares.** Preferred shares typically entitle a holder to a fixed dividend, which is received before any common stockholders may receive dividends. As a general rule, they qualify for inclusion in tier

1 capital if losses are absorbed while the issuer operates as a going concern. Clauses, covenants, and restrictions that make these shares more debit-like may cause them not to be acceptable as tier 1 capital.

– **Retained earnings.** Cumulated earnings due to realized profit not distributed among shareholders are usually assimilated to common shares and included within the tier 1 capital.

- **Subordinated debts.** A few characteristics are required for a debt to be qualified as subordinated and be classified as tier 2 capital. A limited repayment right usually identifies a subordinated debt in the event of default. In more detail, these financial instruments are usually limited to a nonpayment amount. The only remedy afforded to the investor is to petition for the winding up of the bank. In that case, investors are allowed to claim in the winding-up process. Additionally, the maturity needs to be longer than a certain period (e.g., 5 years) and the amortization process is required to be aligned with regulatory schemes.

- **Noncontrolling interests.** A noncontrolling interest is created when a depository institution owns a controlling interest in but not 100% of a subsidiary. The remaining interest is owned by third parties, referred to as *noncontrolling shareholders*. The noncontrolling interest should absorb losses in the subsidiary commensurate with the subsidiary's capital needs. It should not represent an essentially risk-free or low-risk investment for the holders of the subsidiary capital instrument.

2. **Computation of deductions.** As per the regulatory capital computation process, deductions are estimated as detailed below:

- **Goodwill and intangible assets.** Goodwill and intangibles are usually included among assets, according to accounting prescriptions. However, their intangible nature dilutes equity's capability to absorb losses. For this reason a 100% deduction from tier 1 is required.

- **Shortfall (provisions excess).** According to the BIS (2004):

in order to determine provision excesses or shortfalls, banks will need to compare the IRB measurement of expected losses EAD [exposure at default] × PD [probability of default] × LGD [loss given default] with the total amount of provisions that they have made, including both general, specific, portfolio-specific general provisions as well as eligible credit revaluation reserves discussed above. As previously mentioned, provisions or write-offs for equity exposures will not be included in this calculation. For any individual bank, this comparison will produce a "shortfall" if the expected loss exceeds the total provision, or an excess if the total provision exceeds the expected loss.

The idea behind this deduction is outlined through Fig. 6.2. As described in Chapter 4, the regulatory capital faces unexpected losses.

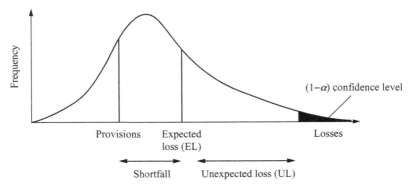

FIG. 6.2 Shortfall mechanics.

Indeed, expected losses are supposed to be captured by provisions and deducted from equity. Therefore the difference (shortfall) between expected losses and provisions is deducted from capital.

Example 6.1 details how to compute the shortfall.

Example 6.1 Shortfall

A bank has $8.50 billion of tier 1 capital and $5.50 billion of tier 2 capital. Its total capital accounts for $14 billion. Table 6.4 summarizes the expected loss (due to Basel IRB parameters) and provisions. A shortfall originates from the difference between these two components.

TABLE 6.4 Credit Shortfall Computed as the Difference Between Basel II Expected Loss and Accounting Provisions ($ Billions)

Portfolio	Expected Loss (Basel II)	Provisions	Shortfall
Performing	10.00	9.00	1.00
Nonperforming	25.00	22.00	3.00
Total	35.00	31.00	4.00

According to Table 6.4, an overall $4 billion shortfall needs to be deducted from capital (for simplicity, no phasing in is considered). Therefore the tier 1 capital reduces to $4.50 billion. Likewise, the total capital drops from $14 billion to $10 billion. The reader is invited to refer to Sections 6.2.2 and 6.3.2 for a comprehensive description of the shortfall mechanics.

The next section focuses on Bank Alpha to show how to implement the above mechanism in practice.

6.2.2 Bank Alpha's Stress Testing Regulatory Capital

Bank Alpha embodies a useful example for an in-depth regulatory capital analysis. Table 6.5 summarizes Bank Alpha's balance sheet items to be part of the regulatory capital. Data refer to t_0 (i.e., stress testing starting point).

The following key assumptions qualify Bank Alpha's regulatory capital.

- **Subordinated debts.** These instruments are assumed to be qualified as tier 2 capital.
- **Noncontrolling interests.** These interests are assumed to share tier 2 capital characteristics.
- **Common shares.** They belong to CET1.
- **Preferred shares.** All these shares are assumed to be classifiable as tier 1 capital.
- **Retained earnings.** They belong to CET1.
- **Goodwill.** Bank Alpha's balance sheet embodies a $1.50 billion goodwill to be deducted from tier 1 capital.
- **Shortfall performing.** Following the guidelines outlined in the previous section, Bank Alpha's expected losses on the performing portfolio are assumed to account for $0.66 billion, with a corresponding $0.60 billion collective provisions. The shortfall is $0.06 billion.
- **Shortfall nonperforming.** According to the BIS (2006), *the capital requirement (K) for a defaulted exposure is equal to the greater of zero and the difference between its LGD [loss given default] and the bank's best estimate of the expected loss. The risk-weighted asset amount for the defaulted exposure is the product of K, 12.5, and the EAD [exposure at default].*

Therefore the difference between the expected loss and provisioning needs to be additionally compared against the so-called best estimate of the expected loss as detailed in Section 6.3.2.

TABLE 6.5 Bank Alpha's Subordinated Debts, Noncontrolling Interests, and Shareholder Equity at t_0 ($ Billions)

		t_0
Subordinated debts		4.00
Noncontrolling interests		2.00
Shareholder equity		7.00
	Common shares	4.00	
	Preferred shares	1.00	
	Retained earnings	2.00	

TABLE 6.6 Bank Alpha's Regulatory Capital From t_0 to t_3 ($ Billions)

		t_0	CET1	Additional Tier 1 Capital	Tier 2 Capital
Subordinated debts		4.00			4.00
Noncontrolling interests		2.00			2.00
Shareholder equity		7.00			
	Common shares	4.00	4.00		
	Preferred shares	1.00		1.00	
	Retained earnings	2.00	2.00		
Deductions					
	Goodwill		−1.50		
	Shortfall		−0.06		
Regulatory capital			4.44	1.00	6.00

TABLE 6.6 Bank Alpha's Regulatory Capital From t_0 to t_3 ($ Billions)—cont'd

t_1			CET1	Additional Tier 1 Capital	Tier 2 Capital
Subordinated debts		4.00			4.00
Noncontrolling interests		2.00			2.00
Shareholder equity		6.18			
	Common shares	4.00	4.00		
	Preferred shares	1.00		1.00	
	Retained earnings	1.18	1.18		
Deductions	Goodwill		−1.45		
Deductions	Shortfall		−0.16		
Regulatory capital		3.57	3.57	1.00	6.00

(Continued)

TABLE 6.6 Bank Alpha's Regulatory Capital From t_0 to t_3 ($ Billions)—cont'd

t_2			CET1	Additional tier 1 capital	Tier 2 capital
Subordinated debts		4.00			4.00
Noncontrolling interests		2.00			2.00
Shareholder equity		6.45			
	Common shares	4.00	4.00		
	Preferred shares	1.00		1.00	
	Retained earnings	1.45	1.45		
Deductions	Goodwill		−1.40		
Deductions	Shortfall		−0.26		
Regulatory capital			3.79	1.00	6.00

TABLE 6.6 Bank Alpha's Regulatory Capital From t_0 to t_3 ($ Billions)—cont'd

t_3			CET1	Additional Tier 1 Capital	Tier 2 Capital
Subordinated debts		4.00			4.00
Noncontrolling interests		2.00			2.00
Shareholder equity		7.15			
	Common shares	4.00	4.00		
	Preferred shares	1.00		1.00	
	Retained earnings	2.15	2.15		
Deductions	Goodwill		−1.35		
Deductions	Shortfall		−0.36		
Regulatory capital			4.44	1.00	6.00

Table 6.6 summarizes Bank Alpha's regulatory capital starting from t_0 until the end of the stress testing process (i.e., t_3). The reader is invited to verify the alignment of these figures with the balance sheet projection described in Chapter 5.

The next section is devoted to RWA analysis. Capital ratios will be estimated accordingly.

6.3 RISK-WEIGHTED ASSETS AND CAPITAL RATIOS

An intricate muddle of rules presides over the RWA computation. An in-depth investigation of each norm would prevent us from showing the key features of a stress testing exercise. For this reason, the focus narrows down to Bank Alpha. This example compromises accuracy and a broad view of the process. Therefore originating from the macroeconomic scenario introduced in Chapter 2 and the balance sheet structure drawn in Chapter 3, we use Bank Alpha to explain the entire RWA estimation mechanism. In more detail, one needs to consider a stress testing transmission framework to shock risk parameters (see Chapters 3 and 4). Afterward, balance sheet and profit and loss projections are made (as detailed in Chapters 4 and 5). Finally, RWA silos are projected and an aggregation system is applied.

6.3.1 Bank Alpha's Risk-Weighted Assets (From a Silo Perspective)

Market, credit, and operational risk methods are used in combination with balance sheet and profit and loss projections to assess Bank Alpha's stress testing RWAs. A summary of each silo RWA is shown to grasp the key features one needs to bear in mind to compute a bank's overall RWAs under stress.

The following advanced formula is used to compute the RWAs for market risk:

$$RWA_{mkt} = 12.5(VaR_{reg} + SVaR_{reg} + IRC + SSRC), \tag{6.3}$$

where the subscript mkt stands for market, VaR_{reg} is the regulatory value at risk, $SVaR_{reg}$ is the stressed value at risk, IRC stands for incremental risk charge due to the counterpart credit risk. $SSRC$ embraces the risk of loss from changes in the market value of a position that could result from factors other than market movements and includes event risk, default risk, and idiosyncratic risk.

Originating from the advanced approach described in Chapter 3, Table 6.7 highlights Bank Alpha's RWAs for market risk over the 3-year stress testing period. Stressed market risk parameters in conjunction with a small increase of the portfolio lead to a moderate RWA boost. It is worth mentioning that a dynamic balance sheet perspective implies the projection of exposures and changes in volatility due to adverse conditions over time (FRB, 2015). In contrast, a static balance sheet method anchors to the initial portfolio on which stressed parameters are applied. Table 6.7 reflects a dynamic standpoint.

TABLE 6.7 Bank Alpha's Market Risk-Weighted Assets ($ Billions)

	RWAs t_0	RWAs t_1	RWAs t_2	RWAs t_3
VaR	0.30	0.42	0.41	0.40
SVaR	1.13	1.30	1.27	1.26
IRC	0.75	0.88	0.85	0.84
SSRC	0.13	0.16	0.15	0.15
Total	2.31	2.76	2.68	2.65

IRC, Incremental risk charge; RWAs, risk-weighted assets; SSRC, standard specific risk charge; SVaR, stressed value at risk; VaR, value at risk.

For the credit risk area, Table 6.8 summarizes the key features of Bank Alpha's RWAs. The performing portfolio RWAs are computed according to the IRB advanced formula:

$$RWA_{cr} = 12.5 \cdot 1.06 \sum_{i=1}^{n} A_i LGD_i \Phi\left(\frac{\Phi^{-1}(PD_i) + \sqrt{\rho}\Phi^{-1}(0.999)}{\sqrt{1-\rho}} - PD_i\right) adj(M_i),$$

(6.4)

where the subscript cr stands for credit, while i identifies a debtor belonging to the (performing) portfolio,[1] $\Phi^{-1}(PD_i)$ is the standard normal inverse cumulative distribution function, PD_i is the probability of default, LGD_i is the loss given default, and A_i denotes the asset exposure (i.e., exposure at default). As described in Chapter 4, correlation (ρ) and maturity adjustment ($adj(M)$) vary according to the type of exposure.

Credit risk parameters are stressed considering the adverse scenario outlined in Chapter 2. The method explained in Chapter 4 is applied as a bridge to generate the shock. Balance sheet movements are estimated according to the schemes illustrated in Chapter 5. Table 6.8 highlights that adverse macroeconomic conditions cause a severe increase in RWAs for credit risk. The underlying risk parameters hike during the first year and partially smooth down at t_2 and t_3. All in all, an approximate 66% jump is recorded at t_1 compared with t_0. This spike is partially absorbed during the following years.

The nonperforming portfolio is analyzed in Section 6.3.2, where a comparison between expected loss (IRB based) and the best estimate of the expected loss is conducted. This test will allow us to decide whether to record additional RWAs or reduce the regulatory capital.

---

1. For simplicity, the subscript s used in Chapter 4 to specify the sector to which the customer belongs is omitted.

TABLE 6.8 Bank Alpha's Credit Risk-Weighted Assets ($ Billions)

	RWAs t_0	RWAs t_1	RWAs t_2	RWAs t_3
Corporate	14.69	29.38	25.85	23.27
Retail	26.90	40.35	36.32	34.50
Others	13.21	21.14	19.02	17.50
Total	54.80	90.87	81.19	75.27

RWAs, Risk-weighted assets.

As detailed in Chapter 5, Bank Alpha uses the standardized approach to estimate its operational risk. Risk weights ranging from 12% to 18% (denoted as β_{bl}) are applied on the gross income spread among regulatory business lines as detailed below;

$$\text{RWA}_{\text{or,SA}} = 12.5 \frac{\sum_{t=1}^{3} \max(\sum_{bl=1}^{8} GI_{bl,t}\beta_{bl}, 0)}{3}, \tag{6.5}$$

where the subscript *SA* stands for standardized approach and $GI_{bl,t}$ is the gross income of year *t* for business line *bl*. It is worth noting that t refers to the previous three years.

Table 6.9 shows a feeble reduction of RWAs for operational risk during the 3-year stress testing exercise.

TABLE 6.9 Bank Alpha's Operational Risk-Weighted Assets ($ Billions)

	RWAs t_0	RWAs t_1	RWAs t_2	RWAs t_3
Corporate finance	0.11	0.11	0.11	0.10
Trading and sale	0.56	0.55	0.53	0.52
Retail banking	2.40	2.34	2.28	2.22
Commercial banking	4.22	4.11	4.00	3.90
Payment and settlement	0.90	0.88	0.85	0.83
Agency services	0.28	0.27	0.27	0.26
Asset management	0.23	0.22	0.21	0.21
Retail brokerage	0.23	0.22	0.21	0.21
Total	8.93	8.69	8.46	8.24

RWAs, Risk-weighted assets.

The next section targets RWA aggregation. Moving from the regulatory capital described in Section 6.2.2 and the aggregated RWAs of Section 6.3.2, we scrutinize capital ratios in Section 6.3.3.

6.3.2 Risk-Weighted Asset Aggregation

The overall capital requirements computation relies on silo RWAs. However, a more sophisticated mechanism with additional constraints holds. A regulatory floor is the first item marking the difference between a sum of silos and an aggregated measure of risk. The BIS (2015) highlights that *the Basel II framework introduced a capital floor as part of the transitional arrangements for banks using the internal ratings-based (IRB) approach for credit risk and/or an advanced measurement approach (AMA) for operational risk. The objective of the floor was to ensure capital requirements did not fall below a certain percentage of banks' capital requirements under the previous Basel I framework. In July 2009, the Committee agreed to keep in place the Basel I capital floor.*

Moreover, the Basel Committee on Banking Supervision views the role of a capital floor as complementing the leverage ratio introduced as part of Basel III.[2]

As mentioned in Section 6.2, provisions need to be examined together with RWAs and regulatory capital when IRB models are adopted. According to the BIS (2011), the capital requirement for a defaulted exposure is equal to the greater of zero and the difference between its loss given default and the bank's best estimate of the expected loss. The RWA amount for the defaulted exposure is the product of the capital requirement, 12.5, and the exposure at default.

The tenor of the above rules can be further detailed through Example 6.2. The bank introduced in Example 6.1 is further investigated to examine the regulatory floor and the impact of the best estimate of the expected loss on capital ratios.

---

2. The Basel framework prescribes a capital floor based on 80% of the Basel I approach for banks that apply the advanced approaches to calculate capital requirements for credit risk (IRB approach) and operational risk (advanced measurement approach). The US core banks that have exited the parallel run are required to calculate a floor based on 100% of the new US standardized approach. The US agencies have explained that for a typical US bank the US floor will be at least as conservative as the Basel I floor (BIS, 2014a). Article 500 of the EU Capital Requirements Regulation states that banks shall hold own funds that are at all times more than or equal to 80% of the total minimum amount of the own funds that the institution would be required to hold under Article 4 of Directive 93/6/EEC as that directive and Directive 2000/12/EC of the European Parliament and of the Council of Mar. 20, 2000 relating to the taking up and pursuit of the business of credit institutions stood prior to Jan. 1, 2007. The BOE (2013) refers to Article 500 of the EU Capital Requirements Regulation to apply the regulatory floor.

Example 6.2 Credit Risk and Capital Ratios: The Role of the Best Estimate of the Expected Loss

The bank introduced in Example 6.1 has $8.50 billion of tier 1 capital and $5.50 billion of tier 2 capital. As in Example 6.1, the best estimate of the expected loss is assumed to be aligned with the expected loss. Therefore an overall $4 billion capital deduction is recorded as detailed in Table 6.10. As a result, the tier 1 capital reduces to $4.50 billion and the total capital becomes $10 billion.

TABLE 6.10 Shortfall When the Expected Loss Equals the Best Estimate of the Expected Loss ($ Billions)

Portfolio	EL (Basel II)	Provisions	EL$_{BE}$	RWA Nonperforming	Capital Shortfall
Performing	10.00	9.00			1.00
Nonperforming	25.00	22.00	25.00		3.00
Total	35.00	31.00			4.00

EL, Expected loss; EL$_{BE}$, best estimate of the expected loss; RWA, risk-weighted asset.

Let us additionally assume that Basel I credit and market RWAs sum up to $100 billion. By application of the 80% regulatory threshold, a floor is set up at $80 billion. Moreover, the sum of (advanced modeling) the RWAs for market, credit, and operational risk is $75 billion. In this context, the floor being higher than the sum of RWA silos, a $5 billion add-on is needed.

The risk-based capital ratios when the expected loss equals the best estimate of the expected loss are computed as follows:

- Tier 1 capital ratio $\frac{4.50}{80} = 5.625\%$.

- Total capital ratio $\frac{10}{80} = 12.5\%$.

Let us now modify our assumption on the best estimate of the expected loss as detailed in Table 6.11

TABLE 6.11 Shortfall When the Expected Loss is Greater Than the Best Estimate of the Expected Loss ($ Billions)

Portfolio	EL (Basel II)	Provisions	EL$_{BE}$	RWA Nonperforming	Capital Shortfall
Performing	10.00	9.00			1.00
Defaulted	25.00	22.00	22.00	37.50	
Total	35.00	31.00		37.50	1.00

EL, Expected loss; EL$_{BE}$, best estimate of the expected loss; RWA, risk-weighted asset.

Capital deductions account for $1 billion. This causes a tier 1 capital reduction to $7.50 billion and a total capital drop to $13 billion. The difference between the expected loss and the best estimate of the expected loss (i.e., $3 billion) increases the RWAs by $3 billion ×12.5 = $37.50 billion. Therefore the sum of RWAs is $75.00 billion + $37.50 billion = $112.50 billion. In this case, the floor is crossed. No additional add-on is required.

(Continued)

Example 6.2 Credit Risk and Capital Ratios: The Role of the Best Estimate of the Expected Loss—cont'd

Finally, risk-based capital ratios when the expected loss is greater than the best estimate of the expected loss are listed below:

- Tier 1 capital ratio $\frac{7.50}{112.50} = 6.67\%$.

- Total capital ratio $\frac{13.00}{112.50} = 11.55\%$.

  The comparison between capital ratios in these two scenarios shows the importance of the best estimate of the expected loss in the capital assessment process.

The next section summarizes the overall process examined throughout the stress testing and capital requirement journey by means of Bank Alpha's illustrative example.

6.3.3   Bank Alpha's Stress Testing Capital Ratios

Three last steps are required to compute Bank Alpha's capital ratios. Firstly, a shortfall is assessed. Secondly, the aggregated RWAs are estimated on the basis of the Basel I floor. Finally, capital ratios are assessed by use of the results of the previous steps:

- **Shortfall.** As a first step, Table 6.12 summarizes the key components to compute RWAs and capital shortfalls. As detailed in the previous sections, a distinction between performing and defaulted portfolios is made. Thus a comparison between the expected loss, provisions, and best estimate of the expected loss is made. Table 6.12 shows that, for the performing portfolio, the expected loss is greater than or equal to provisions during the entire stress testing exercise (i.e., t_1, t_2, t_3). The performing EL is $0.66 billion at t_0, while provisions account for $0.60 billion. The $0.06 billion difference is the capital shortfall at t_0. This shortfall was highlighted in Table 6.6. For the defaulted portfolio, the best estimate of the expected loss is aligned with provisions. At t_0 the expected loss is greater than or equal to the best estimate of the expected loss. Thus, a $2.50 billion RWA shortfall is computed (i.e., $4.60 billion − $4.40 billion = $0.20 billion × 12.5 = $2.50 billion). The same mechanism is applied during the entire stress testing exercise as detailed in Table 6.12. Provisions are aligned with those illustrated in Chapter 5 (i.e., Bank Alpha's credit portfolio projection).
- **Aggregated RWAs.** The sum of market and credit Basel I RWAs is the starting point to estimate the aggregated RWAs (based on advanced models). A regulatory percentage is applied to Basel I RWAs to assess a floor. Table 6.13 highlights both these components. The second row outlines the floor. Therefore RWAs for market, performing credit, operational, and other risks are shown. Moreover, a nonperforming RWA is considered. The sum of all these components is compared against the floor. The difference between this latter sum and the Basel I floor, if positive, is recorded as add-

TABLE 6.12 Comparison of Bank Alpha's Expected Loss, Provisions, and Best Estimate of the Expected Loss to Compute Risk-Weighted Asset and Capital Shortfalls Along the Stress Testing Horizon ($ Billions)

	Portfolio	EL (Basel II)	Provisions	EL_{BE}	RWA Nonperforming	Capital Shortfall
t_0	Performing	0.66	0.60			0.06
	Nonperforming	4.60	4.40	4.40	2.50	
	Total	5.26	5.00		2.50	0.06
t_1	Performing	0.97	0.85			0.12
	Nonperforming	6.93	6.58	6.58	4.38	
	Total	7.90	7.43		4.38	0.12
t_2	Performing	1.05	0.96			0.09
	Nonperforming	7.87	7.74	7.74	1.63	
	Total	8.92	8.70		1.63	0.09
t_3	Performing	0.99	0.93			0.06
	Nonperforming	8.52	8.46	8.46	0.75	
	Total	9.51	9.39		0.75	0.06

EL, Expected loss; EL_{BE}, best estimate of the expected loss; RWA, risk-weighted asset.

TABLE 6.13 Bank Alpha's Overall risk-Weighted Asset Computation Along the Stress Testing Horizon ($ Billions)

	t_0	t_1	t_2	t_3
Basel I RWAs	95.00	101.00	103.00	104.00
Basel I floor	76.00	80.80	82.40	83.20
Market risk RWAs	2.31	2.76	2.68	2.65
Credit risk RWAs	54.80	90.87	81.19	75.27
Operational risk RWAs	8.93	8.69	8.46	8.24
Other RWAs	1.20	1.30	1.85	1.12
RWA nonperforming	2.50	4.38	1.63	0.75
RWAs (no floor)	69.74	108.00	95.81	88.03
RWA add-on	6.26	–	–	–
Total RWAs	76.00	108.00	95.81	88.03

RWA, Risk-weighted asset.

TABLE 6.14 Bank Alpha's Regulatory Capital ($ Billions), Risk-Weighted Assets ($ Billions), and Capital Ratios (%) During the Stress Testing Exercise

		t_0	t_1	t_2	t_3
Regulatory capital ($ billions)					
	CET1	4.44	3.57	3.79	4.44
	Tier 1	5.44	4.57	4.79	5.44
	Total capital (tier 1 + tier 2)	11.44	10.57	10.79	11.44
RWAs		76.00	108.00	95.81	88.03
Capital ratios (%)					
	Cet1 ratio	5.84	3.31	3.96	5.04
	Tier 1 ratio	7.16	4.23	5.00	6.18
	Total capital (tier 1 + tier 2) ratio	15.05	9.79	11.26	13.00

RWAs, Risk-weighted assets.

on RWAs. Table 6.13 pinpoints a steep increase in the total RWAs at t_1. This is mainly due to the jump of the credit risk RWAs from t_0 to t_1 described in Section 6.3.1.

- **Capital ratios.** Table 6.14 summarizes capital ratios during the stress testing exercise. The adverse scenario at t_1 heads a harsh increase in RWAs accompanied by a dramatic reduction in capital ratios. Therefore starting from a strong capital framework at t_0, the common equity ratio falls below the 4.50% Basel III threshold. Bank Alpha also falls below the tier 1 ratio 6% threshold at t_1 and t_2. This implies that Bank Alpha would probably have to consider a capitalization plan or business restructuring.

It is worth highlighting that, apart from pillar 1, other risks need to be considered to assess bank resilience against adverse conditions. In this regard, Fig. 6.3 summarizes the main components one needs to bear in mind. Common equity, tier 1 capital, and total capital ratios are the very bottom threshold. Then, bank-specific buffers are required to address pillar 2. Systemic requirements as well as stress test-specific issues (i.e., countercyclical and conservation buffers) are also included.

Pillar 2 risks will be considered as part of the integration process described in Chapter 7. In what follows, the attention is devoted to leverage and liquidity ratios.

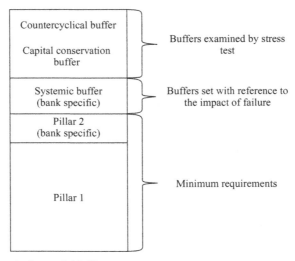

FIG. 6.3 Stress testing capital buffers.

6.4 LEVERAGE AND LIQUIDITY RATIOS

The leverage ratio was introduced after the 2007–09 crisis to serve as a backstop for risk-based capital ratios. The intention was to create a secondary metric that was simple and transparent to assess an appropriate balance sheet size. Moreover, the financial crisis highlighted difficulties in maintaining an adequate balance between cash outflows and inflows. Indeed, unprecedented levels of liquidity support were required from central banks to sustain the financial system. Nonetheless, even with such extensive support a number of banks failed or required resolution. These events were preceded by several years of ample liquidity in the financial system. Liquidity risk and its management did not receive an adequate level of scrutiny and priority as other risk areas. The crisis illustrated how quickly and severely liquidity risks can blow up and funding evaporate. Basel III introduced two additional regulatory standards: the LCR and the NSFR.

6.4.1 Leverage Ratio

The Northern Rock case introduced in Chapter 1 highlights how an excessive leverage may be critical for bank solvency. The leverage ratio can be summarized as a measure of capital as a proportion of total adjusted assets as detailed below:

$$\text{Leverage ratio} = \frac{\text{Capital exposure}}{\text{Exposure measure}}. \tag{6.6}$$

Capital is calculated by use of the tier 1 definition described in the previous sections. In contrast, the financial accounting balance sheet is used as a starting point to measure the exposure. Specific provisions and valuation adjustments may be deducted from the exposure to which they relate. As a general principle, collateral, guarantees, and purchased credit risk mitigation may not be deducted from exposures (BIS, 2014b). A bank's total exposure measure is the sum of the items listed below:

- **On-balance sheet exposures.** All balance sheet assets are included according to their accounting measurement. On-balance sheet derivatives, collateral, and covenants for securities finance transactions, different from those described below, are also included.
- **Derivative exposures.** For derivatives, two types of exposures are considered. On the one hand, exposure may arise from the instrument underlying the derivative. On the other, a counterpart credit risk exposure is taken into account. All in all, derivatives are measured with use of the accounting exposure that reflects the fair value of the contract. An add-on for potential future exposure is also used to ensure a consistent conversion to a loan equivalent amount.
- **Securities finance transactions.** Secured lending and borrowing is an important source of leverage. Repurchase agreements and securities finance are included by use of the accounting measure of exposure. Regulatory netting rules are applied.
- **Off-balance sheet items.** Off-balance sheet items including commitments, letters of credit, failed transactions, and unsettled securities are subject to a uniform 100% credit conversion factor. The only exception is that any commitments that are unconditionally cancelable by the bank at any time without prior notice may have a credit conversion factor of 10%.

During the parallel run period, between 2013 and 2017, a minimum ratio of 3% is tested. The Basel Committee on Banking Supervision will investigate whether this percentage and the design of the ratio are appropriate over a full credit cycle and different types of business models. On the basis of the results for the parallel run period, there might be adjustments in the first half of 2017. The leverage ratio will become an explicit requirement as of Jan. 1, 2018.

In the next section, Bank Alpha's leverage ratio is investigated during the entire stress testing exercise.

6.4.2 Bank Alpha's Stress Testing Leverage

It is worth remarking that a few simplifications were made in Chapter 3 when Bank Alpha was introduced. One of the most important was to assume the absence of derivatives and off-balance sheet operations. Despite the relevance of these instruments in the current economy, their exclusion did not dramatically affect the discussion. Nonetheless, one needs to bear in mind that these components may have an important role when one is assessing the leverage.

In line with Bank Alpha's balance sheet structure, a couple of additional questions arise. On the one hand, one may be puzzled by the inclusion of intangibles among assets. On the other hand, the treatment of shortfall needs to be explained more precisely in terms of its contribution to the numerator of Eq. (6.6). The BIS (2014b) addresses both these questions. Indeed, paragraph 16 states:

> To ensure consistency, balance sheet assets deducted from Tier 1 capital (as set out in paragraphs 66 to 89 of the Basel III framework) may be deducted from the exposure measure. Two examples follow:
>
> • Where a banking, financial or insurance entity is not included in the regulatory scope of consolidation as set out in paragraph 8, the amount of any investment in the capital of that entity that is totally or partially deducted from CET1 capital or from Additional Tier 1 capital of the bank following the corresponding deduction approach in paragraphs 84 to 89 of the Basel III framework may also be deducted from the exposure measure.
> • For banks using the internal ratings-based (IRB) approach to determining capital requirements for credit risk, paragraph 73 of the Basel III framework requires any shortfall in the stock of eligible provisions relative to expected losses to be deducted from CET1 capital. The same amount may be deducted from the exposure measure.

According to the above, Table 6.15 summarizes the components of the leverage ratio and its evolution during the stress testing exercise. The minimum 3% is exceeded during the entire stress testing period.

In the following sections liquidity is examined by our focusing on the LCR and the NSFR.

TABLE 6.15 Bank Alpha's Leverage Ratio ($ Billions)

	t_0	t_1	t_2	t_3
On-balance sheet items	100.00	105.30	109.62	114.18
Assets deducted from tier 1 capital	−1.56	−1.61	−1.66	−1.71
Total balance sheet exposure	98.44	103.69	107.96	112.47
Tier 1 capital	5.44	4.57	4.79	5.44
Leverage ratio (%)	5.53	4.41	4.44	4.84

6.4.3 Liquidity Coverage Ratio

The LCR identifies the amount of high-quality liquid assets that an institution holds to offset the net cash outflows (operational liquidity risk). The LCR assumes an acute short-term, 30-day stress scenario. The specified scenario entails both bank-specific and market-wide shocks built on actual circumstances experienced during the financial crisis.

The quantification of liquid asset requirements and the formulation of qualitative needs is undertaken by consideration of observed trends in the value of assets under stressed conditions. Additionally, the expected and observed behaviors of inflows and outflows during periods of unexpected volatility and crisis need to be taken into account.

One needs to consider their capacity to generate cash through sale or secured borrowing to test the quality of liquid assets . The test is favorable when no loss of value is experienced in periods of severe idiosyncratic and market stress. High-quality assets should ideally be eligible at central banks for intraday liquidity needs and overnight facilities.

The LCR is built in terms of the following proportion:

$$LCR = \frac{\text{High-quality liquid assets}}{\text{Total net cash outflow}}, \qquad (6.7)$$

where the numerator and denominator of Eq. (6.7) are obtained as detailed below:

- **High-quality liquid assets.** This pool comprises readily marketable securities with the potential to generate liquidity under stress. The buffer of high-quality liquid assets comprises level 1 and level 2 assets as detailed below. Level 1 instruments can be included without limit, while a 40% limit (out of the total) is applied to level 2 assets. This nonexhaustive list aims to pinpoint the key items to be considered.
  - **Level 1 assets.** These instruments are included in the buffer at their market value. The following assets are assigned a factor of 100% reflecting a presumed high liquidity under stress:
    * Cash.
    * Central bank reserves able to be drawn down in times of stress.
    * Liquid, marketable securities issued by or guaranteed by sovereign states, central banks, and certain international organizations and that qualify for a 0% risk weight under the Basel II standardized approach for credit risk.
    * Certain nonzero RWAs may also be included where these match an institution's jurisdictional currency liquidity needs or operational requirements.
  - **Level 2 assets.** These assets can be included subject to a minimum 15% supervisory haircut to their market value. More specifically, a distinction

is made between level 2A and level 2B assets, having haircuts of 15% and 50% respectively. In all cases, level 2 assets are capped at 40% (after haircut) of the total buffer. The following list focuses on level 2A items (certain additional assets may be included in level 2 at the discretion of national authorities):

* Liquid, marketable securities issued by or guaranteed by sovereign states, central banks, and certain international organizations and that qualify for a 20% risk weight under the Basel II standardized approach for credit risk.
* Certain corporate bonds (senior status, vanilla) of at least an AA rating or equivalent.
* Covered bonds of at least an AA rating or equivalent.

• **Total net cash outflow.** The denominator of the LCR relies on cumulative outflows deducted from inflows over a 30-day stress period. Inflows are capped at 75% of outflows. A weighting scheme is designed to assign a specific weight (run-off rate) to each outflow and inflow category. In what follows, a nonexhaustive list is shown to highlight the main items to be considered:

– **Outflows**:
* For stable retail deposits, a factor of 5% (run-off rate) is applied.
* For less stable retail deposits, the run-off rate is 10%.
* Unsecured wholesale funding with operational relationships is subject to a 25% run-off rate.

– **Inflows**:
* Maturing reserve repo or securities borrowing transactions secured by level 1 assets are subject to a 0% cash inflow rate. Lines of credit, liquidity facilities, and other contingent funding receive a 0% inflow rate.
* For level 2 collateral, the rate is 15%.
* For retail and small business inflows, the rate is 50%.
* For nonlevel 1 or nonlevel 2 assets, the ratio is 100%.

The next section focuses on Bank Alpha's LCR during the stress testing exercise.

6.4.4 Bank Alpha's Stress Testing Liquidity Coverage Ratio

Bank Alpha is investigated from the regulatory liquidity perspective as detailed in Table 6.16.

Assets and liabilities are assigned a factor according to what was described in Section 6.4.3. More specifically, a factor of 100% is applied to cash and trading account instruments (the latter have a 0% risk weight under the Basel II standardized approach). A 15% haircut (i.e., factor of 85%) is applied to interest-bearing deposits toward high-rating banks. On the total net cash outflow side, a

TABLE 6.16 Bank Alpha's Liquidity Coverage Ratio ($ Billions)

				Factor (%)	Assets/Flow
t_0	A1 assets	Cash	2.00	100.00	2.00
		Trading account	3.00	100.00	3.00
	A2 assets	Interest-bearing deposits	3.00	85.00	2.55
	Total liquid assets				7.55
	Outflows	Retail	45.00	5.00	2.25
		Wholesale	25.00	25.00	6.25
		Total outflows			8.50
	Inflows	Total inflows	2.50	50.00	1.25
	Net outflows				7.25
	LCR (%)				**104.14**
t_1	A1 assets	Cash	2.00	100.00	2.00
		Trading account	3.00	100.00	3.00
	A2 assets	Interest-bearing deposits	3.90	85.00	3.32
	Total liquid assets				8.32
	Outflows	Retail	50.03	5.00	2.50
		Wholesale	25.00	25.00	6.25
		Total outflows			8.75
	Inflows	Total inflows	0.97	50.00	0.48
	Net outflows				8.27
	LCR (%)				**100.60**
t_2	A1 assets	Cash	2.00	100.00	2.00
		Trading account	3.00	100.00	3.00
	A2 assets	Interest-bearing deposits	3.90	85.00	3.32
	Total liquid assets				8.32
	Outflows	Retail	53.31	5.00	2.67
		Wholesale	25.00	25.00	6.25
		Total outflows			8.92
	Inflows	Total inflows	1.66	50.00	0.84
	Net outflows				8.08
	LCR (%)				**102.97**

(Continued)

TABLE 6.16 Bank Alpha's Liquidity Coverage Ratio ($ Billions)—cont'd

				Factor (%)	Assets/Flow
t_3	A1 assets	Cash	2.00	100.00	2.00
		Trading account	3.00	100.00	3.00
	A2 assets	Interest-bearing deposits	3.90	85.00	3.22
	Total liquid assets				8.32
	Outflows	Retail	56.40	5.00	2.82
		Wholesale	25.00	25.00	6.25
		Total outflows			9.07
	Inflows	Total inflows	2.16	50.00	1.08
	Net outflows				7.99
	LCR (%)				**104.13**

LCR, Liquidity coverage ratio.

distinction is made between outflows and inflows. Retail outflows are assigned a factor of 5% because of their stable nature. In contrast, a factor of 25% is applied to wholesale (less stable) deposits. From the inflow standpoint, a factor of 50% is applied because of the retail and small business counterpart nature of the flows. Additionally, the 75% outflow cap is never triggered by inflows.

Bank Alpha highlights a solid liquidity structure mainly due to the nature of its assets and liabilities. The LCR exceeds the 100% threshold during all of the stress testing exercise. Nonetheless, Chapter 7 will show that under some conditions deposit funding instability may undermine bank resilience.

Once completing the LCR analysis, the next section focuses on the net stable funding ratio.

6.4.5 Net Stable Funding Ratio

The NSFR aims to ensure that banks hold a minimum amount of stable funding to run their business. It is based on the liquidity characteristics of their cash outflows and inflows over a 1-year horizon. This ratio has been introduced to reduce maturity mismatches between assets and liabilities, to cover an extended firm-specific stress scenario. In such a context, a bank may encounter a significant decline in its profitability or solvency arising from its risk profile. Additionally, a potential downgrade of its debt or deposit rating may call into question the credit quality of the institution. Under this kind of scenario, borrowings become difficult and usual open market operations may be subject to haircuts.

The NSFR is computed as follows:

$$\text{NSFR} = \frac{\text{Available stable funding}}{\text{Required stable funding}}, \tag{6.8}$$

where items classifiable among available stable funding (ASF) and required stable funding (RSF) are listed below:

- **ASF.** Stable funding is defined as the portion of those types of equity and liability financing expected to provide reliable sources of funds over a 1-year time horizon to cover conditions of extended stress. Firstly, the carrying value of bank equity and liabilities is assigned to one of the following five categories:
  - Total regulatory capital (excluding tier 2 instruments with residual maturity of less than 1 year); other capital instruments and liabilities with effective residual maturity of 1 year or more. The ASF factor is 100%.
  - Stable nonmaturity (demand) deposits and term deposits with residual maturity of less than 1 year provided by retail and small business customers. The ASF factor is 95%.
  - Less stable nonmaturity deposits and term deposits with residual maturity of less than 1 year provided by retail and small business customers. The ASF factor is 90%.
  - Funding with residual maturity of less than 1 year provided by nonfinancial corporate customers; operational deposits and funding with residual maturity of less than 1 year from sovereign states, public sector entities, and multilateral and national development banks; other funding with residual maturity between 6 months and less than 1 year not included in the above categories, including funding provided by central banks and financial institutions. The ASF factor is 50%.
  - All other liabilities and equity categories not included in the above categories have an ASF factor of 0%.

  Then the amount assigned to each category needs to be multiplied by an ASF factor ranging from 0% to 100%. The total ASF is the sum of the weighted amounts.

- **RSF.** The amount of stable funding required by supervisors is measured by taking into account the characteristics of the liquidity risk profiles of assets and off-balance sheet exposures. The RSF is calculated as the sum of the value of assets held, multiplied by a specific factor assigned to each asset type. Some of the major categories are listed below:
  - Coins and banknotes, central bank reserves, and claims on central banks with residual maturities of less than 6 months. The RSF factor is 0%.
  - Unencumbered level 1 assets, excluding coins, banknotes, and central bank reserves. The RSF factor is 5%.
  - Unencumbered loans to financial institutions with residual maturities of less than 6 months, where the loan is secured against level 1 assets

and where the bank has the ability to freely prehypothecate the received collateral for the life of the loan. The RSF factor is 10%.

– All other unencumbered loans to financial institutions with residual maturities of less than 6 months not included in the above categories (unencumbered level 2A assets). The RSF factor is 15%.

– Unencumbered level 2B assets; high-quality liquid assets encumbered for a period of 6 months or more and less than 1 year; loans to financial institutions and central banks with residual maturities between 6 months and less than 1 year; deposits held at other financial institutions for operational purposes; all other assets not included in the above categories with residual maturity of less than 1 year, including loans to nonfinancial corporate clients, loans to retail and small business customers, and loans to sovereign states and public sector entities. The RSF factor is 50%.

– Unencumbered residential mortgages with a residual maturity of 1 year or more and with a risk weight of less than or 35% under the standardized approach to credit risk; other unencumbered loans not included in the above categories, excluding loans to financial institutions, with a residual maturity of 1 year or more and with a risk weight of less than or 35% under the standardized approach. The RSF factor is 65%.

– Cash, securities, or other assets posted as the initial margin for derivative contracts and cash or other assets provided as contributions to the default fund of a clearing house; other unencumbered performing loans with risk weights greater than 35% under the standardized approach and residual maturities of 1 year or more, excluding loans to financial institutions; unencumbered securities that are not in default and do not qualify as high-quality liquid assets with a remaining maturity of 1 year or more and exchange-traded equities; physical traded commodities, including gold. The RSF factor is 85%.

– All other assets not included in the above categories, including nonperforming loans, loans to financial institutions with a residual maturity of 1 year or more, nonexchange-traded equities, and fixed assets; items deducted from regulatory capital, retained interest, insurance assets, subsidiary interests, and defaulted securities. The RSF factor is 100%.

The next section focuses on Bank Alpha's NSFR during the stress testing exercise.

6.4.6 Bank Alpha's Stress Testing Net Stable Funding Ratio

Bank Alpha is investigated from the NSFR perspective as detailed in Table 6.17. Assets and liabilities are assigned a factor according to the details provided in Section 6.4.5. In more detail, a distinction is made between assets and liabilities. On the asset side, a factor of approximately 21% characterizes cash resources during the stress testing exercise. This is due to a mix of cash, central bank reserves, and loans to financial institutions with residual maturity between 6

TABLE 6.17 Bank Alpha's Stress Testing Net Stable funding Ratio Dynamics ($ Billions)

		Exposure	Average Factor (%)	
t_0	Cash resources	8.00	20.63	1.65
	Securities	14.00	100.00	14.00
	Loans	70.00	77.50	54.25
	Other assets	8.00	100.00	8.00
	Total RSF			77.90
	Deposits	70.00	74.64	55.25
	Other liabilities	17.00	26.47	4.50
	Subordinated debts	4.00	75.00	3.00
	Noncontrolling interests	2.00	100.00	2.00
	Shareholder equity	7.00	100.00	7.00
	Total ASF			68.75
	NSFR (%)			**88.25**
t_1	Cash resources	9.00	20.50	1.85
	Securities	15.00	100.00	15.00
	Loans	72.28	77.20	55.80
	Other assets	9.02	100.00	9.02
	Total RSF			81.67
	Deposits	75.04	79.00	59.28
	Other liabilities	18.08	26.05	4.71
	Subordinated debts	4.00	100.00	4.00
	Noncontrolling interests	2.00	100.00	2.00
	Shareholder equity	6.18	100.00	6.18
	Total ASF			76.17
	NSFR (%)			**93.27**
t_2	Cash resources	9.00	21.00	1.89
	Securities	15.00	100.00	15.00
	Loans	75.43	77.20	58.23
	Other assets	10.19	100.00	10.19
	Total RSF			85.31
	Deposits	78.31	77.86	60.97
	Other liabilities	18.86	27.50	5.19
	Subordinated debts	4.00	100.00	4.00

(Continued)

TABLE 6.17 Bank Alpha's Stress Testing Net Stable funding Ratio Dynamics ($ Billions)—cont'd

		Exposure	Average Factor (%)	
	Noncontrolling interests	2.00	100.00	2.00
	Shareholder equity	6.45	100.00	6.45
	Total ASF			78.61
	NSFR (%)			**92.15**
t_3	Cash resources	9.00	21.00	1.89
	Securities	15.00	100.00	15.00
	Loans	78.52	78.05	61.28
	Other assets	11.66	100.00	11.66
	Total RSF			89.83
	Deposits	81.40	77.07	62.73
	Other liabilities	19.63	27.40	5.38
	Subordinated debts	4.00	100.00	4.00
	Noncontrolling interests	2.00	100.00	2.00
	Shareholder equity	7.15	100.00	7.15
	Total ASF			81.26
	NSFR (%)			**90.46**

ASF, Available stable funding; NSFR, net stable funding ratio; RSF, required stable funding.

months and 1 year. Securities and other assets not classifiable among other most favorable exposures are assigned a factor of 100%. Finally, the wide range of loans included in Bank Alpha's balance sheet have a weight oscillating around 78% during the 3-year stress testing period.

On the liability side, a high quota of deposits is classified among stable nonmaturity and less stable nonmaturity deposits. Additionally, funding with residual maturity of less than 1 year provided by nonfinancial customers causes the ASF factor to fluctuate around 78% along the stress testing horizon. A mix of different facilities characterizes other liabilities, to which a factor oscillating around 27% is assigned. $1 billion subordinated debts have a factor of 0% at t_0 due to their expiry during the period $[t_0, t_1]$. In contrast, a 100% weight is applied to all others and from t_1 onward because of their long term maturity. All other equity instruments are assigned a 100% weight.

All in all, Table 6.17 highlights a substantial balance between assets and liabilities. It is also coherent with the asset and liability management liquidity analysis described in Chapter 3. Nonetheless, the assumption of stable deposits is critical for the overall assessment. This will be illustrated in Chapter 7 when we focus on a fully integrated liquidity analysis.

Example 6.3 helps us understand liquidity strategic planning based on the LCR and NSFR.

Example 6.3 LCR and NSFR

The chief executive officer of a bank asks the chief risk officer to define strategies to improve the LCR and NSFR to reach a 100% target. The starting point is the balance sheet shown in Table 6.18.

Table 6.19 shows the weights to be applied to each asset and liability category to compute both the LCR and the NSFR. The 75% outflow limit is applied to inflows. Moreover, all inflows can be deducted from outflows. According to the weighting scheme described in Section 6.4.3, weighted liquid assets amount to $13 million, while net outflows are $59 million (i.e., $62 billion outflows − $3 billion inflows). The LCR is $\frac{13.00}{59.00} = 22.03\%$.

With respect to the NSFR, available stable funds amount to $66.75 million, while requested stable funds amount to $75.75 million. The NSFR is $\frac{66.75}{75.75} = 88.12\%$.

The first strategy considered by the chief risk officer is to deleverage $20 million by selling corporate credit exposures and correspondingly reducing other funding having maturity within the year. This idea is outlined in Table 6.20.

According to Table 6.20, this strategy would allow the bank to improve its NSFR as follows: $\frac{56.75}{55.75} = 102\%$. However the LCR would not be substantially affected: $\frac{13.00}{39.00} = 33\%$. Hence additional changes are required to achieve an LCR of 100% or greater.

The chief risk officer considers an additional increase of stable deposits to finance high-quality, unencumbered liquid assets (e.g., bank deposits) as detailed in Table 6.21

During the meeting with the chief executive officer, the chief risk officer presents the following two-step proposal:

- Deleverage by selling $30 million of credits and correspondingly reduce short-term funds (the original strategy was based on $20 million).
- Stable deposits increase by $17 million to finance bank deposits.

This strategy would allow the bank to improve its NSFR ($\frac{67.90}{45.75} = 148\%$) and hit the LCR target ($\frac{30.00}{29.85} = 101\%$).

6.5 SUMMARY

An introduction to the definition of own funds paved the way to understanding how to compute capital ratios. The distinction between tier 1 common equity, additional tier 1 capital, and tier 2 capital was described by our highlighting the key differences between accounting and regulatory capital. As part of the denominator of capital ratios, market, credit, and operational RWAs were examined from a silo perspective. Then they were aggregated to compute the overall

TABLE 6.18 Bank's Balance Sheet ($ Millions)

Assets			
Cash resources			8.00
	Cash	5.00	
	Central bank reserves	3.00	
Securities			17.00
	Securities (0% RWAs)	10.00	
	Corporate bonds >1 year	7.00	
Loans			75.00
	Corporate credits >1 year	30.00	
	Loan portfolio corporate <1 year	5.00	
	Loan portfolio retail < 1year	25.00	
	Mortgage loans (35% RWAs)	15.00	
Total assets			100.00
Liabilities			
Deposits			25.00
	Stable deposits	20.00	
	Less stable deposits	5.00	
Other liabilities			55.00
	Corporate operational	10.00	
	Bonds > 1 year	10.00	
	Other funding < 1 year	35.00	
Subordinated debts			10.00
Noncontrolling interests			5.00
Shareholder equity			5.00
Total liabilities			100.00

RWAs, Risk-weighted assets.

bank RWAs by our considering regulatory constraints (e.g., floor, shortfall, and so on). The illustrative example of Bank Alpha was useful to investigate how regulatory rules impact the RWAs. A specific focus targeted the credit risk assessment by comparing the expected loss, provisions, and best estimate of the expected loss. The computation of capital ratios finalized the stress testing process illustrated throughout the book. Then, leverage and liquidity ratios highlighted the importance of monitoring additional nonrisk-based measures to

TABLE 6.19 Liquidity Coverage Ratio and Net Stable Funding Ratio Analysis ($ Millions)

Assets		**Balance**	**LCR Assets**	**Inflows**	**RSFs**
Cash resources		8.00			
	Cash	5.00	100.00%		0.00%
	Central bank reserves	3.00	100.00%		0.00%
Securities		17.00			
	Securities (0% RWAs)	5.00	100.00%		5.00%
	Corporate bonds > 1 year	12.00	0.00%		100.00%
Loans		75.00			
	Corporate credits > 1 year	30.00	0.00%		100.00%
	Loan portfolio corporate < 1 year	5.00		1.00	50.00%
	Loan portfolio retail < 1 year	25.00		2.00	85.00%
	Mortgage loans (35% RWAs)	15.00	0.00%		65.00%
Total Assets		**100.00**	**13.00**	**3.00**	**75.75**

Liabilities		**Balance**	**LCR outflows**		**ASFs**
Deposits		10.00			
	Stable deposits	5.00	5.00%		95.00%
	Less stable deposits	5.00	10.00%		90.00%
Other liabilities		70.00			
	Corporate operational	25.00	25.00%		50.00%
	Bonds >1 year	5.00	100.00%		100.00%
	Other funding <1 year	40.00	100.00%		50.00%
Subordinated debts		10.00	100.00%		100.00%
Noncontrolling interests		5.00	0.00%		100.00%
Shareholder equity		5.00	0.00%		100.00%
Total liabilities		100.00	62.00		66.75

ASFs, Available stable funds; LCR, liquidity coverage ratio; RSFs, required stable funds; RWAs, risk-weighted assets.

TABLE 6.20 Deleveraging Strategy ($ Millions)

Assets					
		Balance	**LCR Assets**	**Inflows**	**RSFs**
Cash resources		8.00			
	Cash	5.00	100.00%		0.00%
	Central bank reserves	3.00	100.00%		0.00%
Securities		17.00			
	Securities (0% RWAs)	5.00	100.00%		5.00%
	Corporate bonds > 1 year	12.00	0.00%		100.00%
Loans		55.00			
	Corporate credits > 1 year	10.00	0.00%		100.00%
	Loan portfolio corporate < 1 year	5.00		1.00	50.00%
	Loan portfolio retail < 1 year	25.00		2.00	85.00%
	Mortgage loans (35% RWAs)	15.00	0.00%		65.00%
Total assets			80.00	13.00	3.00	55.75

Liabilities				
		Balance	**LCR Outflows**	**ASFs**
Deposits		10.00		
	Stable deposits	5.00	5.00%	95.00%
	Less stable deposits	5.00	10.00%	90.00%
Other liabilities		50.00		
	Corporate operational	25.00	25.00%	50.00%
	Bonds >1 year	5.00	100.00%	100.00%
	Other funding <1 year	20.00	100.00%	50.00%
Subordinated debts		10.00	100.00%	100.00%
Noncontrolling interests		5.00	0.00%	100.00%
Shareholder equity		5.00	0.00%	100.00%
Total liabilities		80.00	42.00	56.75

ASFs, Available stable funds; LCR, liquidity coverage ratio; RSFs, required stable funds; RWAs, risk-weighted assets.

TABLE 6.21 Deleveraging and Deposit Increase Strategy ($ Millions)

Assets		Balance		LCR Assets	Inflows	RSFs
Cash resources			25.00			
	Cash	22.00		100.00%		0.00%
	Central bank reserves	3.00		100.00%		0.00%
Securities			17.00			
	Securities (0% RWAs)	5.00		100.00%		5.00%
	Corporate bonds >1 year	12.00		0.00%		100.00%
Loans			45.00			
	Corporate credits >1 year	–		0.00%		100.00%
	Loan portfolio corporate <1 year	5.00			1.00	50.00%
	Loan portfolio retail <1 year	25.00			2.00	85.00%
	Mortgage loans (35% RWAs)	15.00		0.00%		65.00%
Total assets			87.00	30.00	3.00	45.75

Liabilities		Balance	LCR Outflows		ASFs
Deposits		27.00			
	Stable deposits	22.00	5.00%		95.00%
	Less stable deposits	5.00	10.00%		90.00%
Other liabilities		40.00			
	Corporate operational	25.00	25.00%		50.00%
	Bonds >1 year	5.00	100.00%		100.00%
	Other funding <1 year	10.00	100.00%		50.00%
Subordinated debts		10.00	100.00%		100.00%
Noncontrolling interests		5.00	0.00%		100.00%
Shareholder equity		5.00	0.00%		100.00%
Total liabilities		87.00	32.85		67.90

ASFs, Available stable funds; LCR, liquidity coverage ratio; RSFs, required stable funds; RWAs, risk-weighted assets.

ensure a bank is solvent both in the long run and in the short run. A structural balance between own resources and investment together with equilibrated cash flow mismatches finalized the regulatory stress testing analysis.

SUGGESTIONS FOR FURTHER READING

A regulatory perspective inspired this chapter. Therefore the key references relate to the Basel Accord and its methodological notes. The reader may find it useful to read BIS (2011, 2014a,b) and the most recent consultation papers where potential changes to the current regulatory framework are under discussion (BIS, 2015).

EXERCISES

Exercise 6.1 Let us consider the balance sheet shown in Table 6.22. A capital planning exercise is required aimed at improving the core capital ratio by 1.5% and the total capital ratio by 2.0% in the next 2 years.

The bank relies on advanced methods to compute RWAs for market and credit risk, while for the operational risk a basic approach is used. The following simplistic assumptions characterize the computation.

- **Market risk.** The value at risk may be computed by use of the following synthetic measures of volatility: shares have a 0.05% daily variance, while for bonds it is 0.01%. Their covariance is 0.02%.

TABLE 6.22 Balance Sheet at t_0 ($ Billions)

Assets				Liabilities		
Cash resources			2.00	Deposits		60.00
Securities			8.00	Bonds		18.00
	Shares	3.00		Subordinated debts		15.00
	Bonds	5.00		Shareholder equity		7.00
Loans			90.00		Common shares	3.00
	SME	35.00			Preferred shares	1.00
	Corporate	55.00			Retained earnings	3.00
Total assets			100.00	Total liabilities		100.00

SME, Small and medium-sized enterprise.

- **Credit risk.** For small- and medium-sized enterprises the average probability of default is 4%, while the loss given default is 25%. In contrast, the average corporate probability of default is 2% and the loss given default is 20%.
- **Operational risk.** The gross income for the previous 3 years expressed in billion dollars is 0.80, 0.70, and −0.10.

The reader is invited to adopt other suitable assumptions required to perform the capital planning exercise.

Exercise 6.2 Let us consider the balance sheet structure of the bank outlined in Exercise 6.1. A coherent liquidity planning is required to achieve the 100% minimum LCR and NSFR based on the t_0 factors as detailed in Table 6.23.

Solutions are available at www.tizianobellini.com.

TABLE 6.23 Liquidity Coverage Ratio and Net Stable Funding Ratio Factors at t_0

		Assets		
		LCR Assets (%)	Inflows	RSFs (%)
Cash resources		100.00		0.00
Securities				
	Shares	0.00		100.00
	Bonds	100.00		100.00
Loans				
	SME		5.00	100.00
	Corporate		2.00	100.00
		Liabilities		
		LCR Outflows (%)		ASFs (%)
Deposits		10.00		90.00
Bonds		100.00		100.00
Subordinated debts		100.00		100.00
Shareholder equity				
	Common shares	0.00		100.00
	Preferred shares	0.00		100.00
	Retained earnings	0.00		100.00

ASFs, Available stable funds; LCR, liquidity coverage ratio; RSFs, required stable funds; SME, small- and medium-sized enterprise.

REFERENCES

BIS, 1988. International Convergence of Capital Measurement and Capital Standards. Bank for International Settlements, Basel.

BIS, 2004. Modifications to the capital treatment for expected and unexpected credit losses in the new Basel accord. Bank for International Settlements, Basel.

BIS, 2006. Basel II International Convergence of Capital Measurement and Capital Standards: A Revised Framework, Bank for International Settlements, Basel.

BIS, 2011. Basel III: A global regulatory framework for more resilient banks and banking systems. Bank for International Settlements, Basel.

BIS, 2014a. Regulatory Consistency Assessment Programme (RCAP), Assessment of Basel III regulations—United States of America. Bank for International Settlements, Basel.

BIS, 2014b. Revised Basel III leverage ratio framework and disclosure requirements. Bank for International Settlements, Basel.

BIS, 2015. Consultative document: Capital floors: the design of a framework based on standardised approaches. Bank for International Settlements, Basel.

BOE, 2013. The Basel I Floor Supervisory Statement, SS8/13, London.

FRB, 2015. Comprehensive Capital Analysis and Review 2015: assessment framework and results Board of Governors of the Federal Reserve System, Washington, DC.

Chapter 7

Risk Integration

Chapter Outline

Growing interest on risk integration characterizes banking practice. In this regard, a distinction arises between the process to assess long-term bank solvency (i.e., economic capital) and short-term capability to overcome a financial distress (i.e., liquidity mismatch). When one focuses on the economic capital (i.e., the amount of capital that a bank needs to stay solvent), the choice of the risks to integrate is vital. Firstly, different time horizons undermine the theoretical architecture. Additionally, the interaction between assets and liabilities complicates the process. In line with the current literature and banking practice, this chapter focuses mainly on market, interest rate, and credit risks. Other sources of losses are investigated as additional components of a wider integration frame encompassing a 1-year or multiperiod time horizon.

Two main perspectives are followed in our estimating an integrated measure of economic capital. A top-down approach relies on the aggregation of risk silos. In contrast, a bottom-up modeling is based on a micro-level investigation of risk interdependencies. Following the first approach, it is shown how to join independently estimated risks and use all the stress test tools developed throughout the book to assess integrated economic capital.

The last part of this chapter highlights the role of liquidity in banking failures. The focus of the method is to represent the interactions between macroeconomic conditions and the risk of insolvency due to liquidity short-comings.

KEY ABBREVIATIONS AND SYMBOLS

$ALiq_q$ asset liquidity inflows for period $q \leq h$

$ALoss$ asset loss corresponding to $-\Delta A$

A_t asset value at time t

ΔA total asset value change

$CLoss_h$ loss due to defaults for period h

EC_{ES} economic capital, *expected shortfall*

EC_{VaR} economic capital, *value at risk*

E_t equity value at time t

H_i liquidity haircut for obligor i

I_h income for period h

$LLiq_q$ liability liquidity outflows for period $q \leq h$

LM liquidity mismatching

$Loss_h$ total loss for period h

L_t value of liabilities at time t

ΔL total liability value change

$Mloss_h$ market loss for period h

NI_h net interest for period h

$Oloss_h$ operational loss for period h

PD_i probability of default for obligor i

PNI_h performing net interest for period h

PV present value

$R_{d,t}$ interest rate for node d at time t

$\mathbf{x}_t$ macroeconomic variables at time t

$\mathbf{x}_\Delta$ shocked macroeconomic variables

$\mathbb{1}_{i,def}$ default indicator function for obligor i

Ψ_s creditworthiness index for sector s

ϱ_t liquidity shrinkage at time t

7.1 INTRODUCTION

As described in Chapter 6, a modular approach is usually adopted to compute the regulatory capital. Market, credit, and operational risks are added up to obtain pillar 1 risk-weighed assets (RWAs). Nevertheless, one needs to bear in mind that a portfolio is exposed to different imperfectly correlated risks. The recent crisis showed that interdependencies need to be captured. Then a question arises: Is the sum of capital buffers set against these risks smaller than the sum of individual risks? Breuer et al. (2010) discussed this problem by examining the joint distribution of market and credit risk for the banking book and the trading book. They showed that in some cases the joint risk measure may be higher than the sum of the independent risk measures. In some other circumstances the opposite holds.

A series of issues arise when one tries to integrate risks. Firstly, they may have different holding periods. Credit risk, for example, is usually expressed over a 1-year horizon. In contrast, market risk is regularly measured for a 10-day period. Additionally, one may think of integrating these risks by starting from silos (e.g., market, credit, operational, and so on). However, additional interactions may be captured when one considers micro-level interdependencies. The framework developed for stress testing may be very useful in the latter case. All in all, the approach based on silos is commonly indicated as a *top-down approach*, while the method focusing on micro components is denominated a *bottom-up approach*.

Section 7.2 focuses on a top-down perspective. A basic approach relying on a crude aggregation of precalculated risk measures is the forerunner of more sophisticated systems (Dimakos and Aas, 2004). In this regard, three subapproaches are considered. Apart from the above-mentioned basic integration, top-level and base-level methods can be followed. A top-level integration enriches the process by use of a random generator that takes into account the interdependencies among risks (e.g., market, credit, and operational). An additional enhancement characterizes the base-level approach, where scenarios are simulated by use of specific factors (Aas et al., 2007). Some key issues on factor interdependencies as well as model parameter estimates remain unresolved. However, this methodology paves the way to a more comprehensive framework based on bottom-up integration.

The most recent risk management literature highlights the advantages of starting from elementary components to build an integrated framework. A few differences qualify this field of research in terms of the risks included in the study and balance sheet components entering the scope. In particular, when focusing on an economic capital perspective (i.e., the capital that a bank needs to stay solvent), Section 7.3 highlights that market, interest rate, and credit risks are commonly scrutinized. However, in some cases the focus is on assets only, in others it is on both sides of the balance sheet, and in some circumstances also the profit and loss is investigated (a separate reasoning applies to the liquidity analysis). In more detail, the method developed by Grundke (2009) and Grundke (2010) relies heavily on a CreditMetrics-like framework where credit, market, and other minor risks are jointly studied. Alessandri and Drehmann (2010) and Drehmann et al. (2010) concentrate on the asset and liability structure by paying specific attention to the impact of macroeconomic scenarios on income fluctuations. Kretzschmar et al. (2010) use a factor generator to assess the impact of a full fair value asset evaluation, while they model liabilities by matching the balance sheet dynamic. On this subject, Section 7.3 describes a comprehensive economic capital risk integration modeling that uses a method that embraces all the above contributions. In line with Bellini (2013), a framework is proposed where assets and liabilities are jointly considered to integrate market, interest rate, and credit risks. Attention is devoted to both asset and liability present value movements under different macroeconomic scenarios. Additionally, liquidity is investigated in a coherent framework as detailed in Section 7.4. In this regard, the difference between a long- and a short-term bank solvency perspective is shown. On the one hand, the analysis of the economic capital focuses on the structural capability of a bank to face adverse conditions. On the other hand, the study of liquidity risk focuses on a closer horizon.

All the techniques examined throughout the book are used in this chapter. More specifically, starting from the macroeconomic scenario simulation described in Chapter 2, we combine the asset and liability techniques developed in Chapter 3 with the methods studied in Chapters 4–6. All in all, the stress testing mechanism is further enhanced to capture all interconnections among risks threatening bank solvency.

7.2 TOP-DOWN RISK INTEGRATION MODELING

One of the most intuitive ways to integrate risks is to follow a top-down approach. Risk silos are estimated and then joined together. On this subject, a few different approaches have been developed in the literature. In what follows a distinction is made among a basic approach, top-level integration, and base-level integration. In the first case, risk silos are estimated and an integration function (e.g., copula) is used to merge them. In contrast, in the top-level framework each risk is individually estimated on the basis of a conditional realization setting. Subsequently, correlated losses are summed up to obtain the integrated loss, on which a synthetic indicator (e.g., value at risk) is applied. Finally, risk factors are jointly simulated in the base-level integration approach. A transmission mechanism is used to connect factors and risk categories (e.g., credit, operational, market). Consequently, for each of the latter, a loss is computed. A synthetic risk integration indicator is estimated on the basis of their sum. The following sections detail each of these three approaches.

7.2.1 Basic Integration

Following a pillar 1 perspective, a bank's total loss may be represented as the sum of market, credit, and operational risks as follows:

$$Loss_{bi,h} = CLoss_{bi,h} + OLoss_{bi,h} + MLoss_{bi,h}, \tag{7.1}$$

where $Loss_{bi,h}$ is the total loss estimated according to a top-down basic integration (bi) approach over a given horizon (h). The total loss is the sum of credit ($CLoss$), operational ($OLoss$), and market ($MLoss$) losses. The analysis can be extended to other risk sources. However, for simplicity, as a starting point, it is useful to focus on pillar 1 risks only. According to Chapter 6, some additional constraints are imposed by regulators to compute aggregated RWAs. However, credit, operational, and market risk silos are added up in the Basel framework without consideration of their potential interdependencies.

Since the distributions underlying different risks do not necessarily follow the same form, the most effective way to convolve credit, operational, and market risk is to perform a simulation or numerical integration. A normal distribution may be used as in Kuritzkes et al. (2002). Copulas may also be fit for purpose. In all cases, one needs to bear in mind that credit and operational risks are usually computed over a 1-year horizon. In contrast, market risk is typically measured on a daily basis. A unique holding period is required for their integration. Therefore the 1-year horizon seems to be a reasonable compromise. Indeed, it corresponds to the internal capital allocation and budgeting cycle. Moreover, it is the paradigm on which the Basel II Accord relies to compute the regulatory capital.

The standard method for estimating the aggregated economic capital is the variance-covariance method introduced in Chapter 3. In this case, the following holds:

$$Loss_{VaR,bi,(1-\alpha)} = \sqrt{\sum_{r,s} \rho_{r,s} VaR_{r,(1-\alpha)} VaR_{s,(1-\alpha)}}, \qquad (7.2)$$

where $VaR_{r,(1-\alpha)}$ represents the value at risk for the risk r (e.g., market), computed as $VaR_{r,(1-\alpha)} = w_r \sigma_r F_r^{-1}(1-\alpha)$. On the latter, w_r is the portfolio weight of the specific risk source, σ_r is its volatility, and $F_r^{-1}(1-\alpha)$ is the $(1-\alpha)$ quantile of the standardized loss. The same idea applies to $VaR_{s,(1-\alpha)}$ referred to another risk source (e.g., credit).

The above technique relies on the implicit constraint that the distribution of the portfolio corresponds to the marginal ones. This does not necessarily hold in practice. Therefore a copula approach may be used to obtain an integrated loss function starting from the simulation of the marginals (Fig. 7.1). Correlation parameters may be estimated by use of historical time series of credit, operational, and market risks (e.g., RWAs).

The major advantage of this basic approach is its simplicity. In contrast, its inability to capture the underlying risk interactions is its main drawback. For this reason more comprehensive approaches have been developed, as detailed in the following sections.

7.2.2 Top-Level Integration

Fig. 7.2 summarizes the key modeling components of a top-level integrated approach. Credit, operational, and market risk losses rely on randomly generated factors. The simulation is based on a given correlation structure (e.g., multivariate normal, copula, and so on) capturing interdependencies among each risk source. As an intermediate output, each risk loss distribution is computed individually. Finally, all these estimates are added to obtain the total loss.

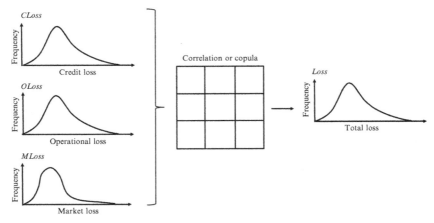

FIG. 7.1 Top-down risk integration: basic integration.

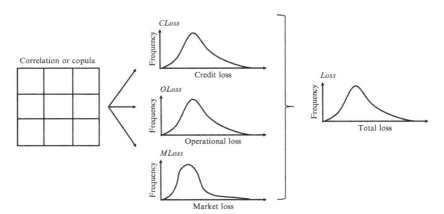

FIG. 7.2 Top-down risk integration: top-level approach.

Dimakos and Aas (2004) start from the multiplication law of probability paradigm. The joint likelihood of credit, operational, and market losses can be decomposed as follows:

$$P(Loss_{tl,h}) = P(CLoss_{tl,h})P(OLoss_{tl,h}|CLoss_{tl,h})P(MLoss_{tl,h}|CLoss_{tl,h}, OLoss_{tl,h}),$$
(7.3)

where the subscript tl stands for top level and the holding period h commonly corresponds to 1-year.

According to Eq. (7.3), the credit risk is taken as the base (to which all the others refer). This is because the credit risk is usually dominant for a bank operating in the commercial business sectors. Starting from the latter statement, one may additionally assume that

$$P(MLoss_{tl,h}|CLoss_{tl,h}, OLoss_{tl,h}) = P(MLoss_{tl,h}|CLoss_{tl,h}), \qquad (7.4)$$

which implies

$$P(Loss_{tl,h}) = P(CLoss_{tl,h})P(OLoss_{tl,h}|CLoss_{tl,h})P(MLoss_{tl,h}|CLoss_{tl,h}). \quad (7.5)$$

As a consequence, if the probability distributions on the right-hand side can be simulated, then the joint distribution may be obtained accordingly. On this subject, Dimakos and Aas (2004) create their system starting from a copula structure to capture the interdependencies between credit and operational risks. Then they make some assumptions on the distributional form of credit, operational, and market risks. Finally, each loss category is summed up to obtain an integrated loss. A synthetic measure such as the value at risk or the expected shortfall summarizes the integrated risk. In what follows, these key steps are analyzed in more detail:

- **Copula simulation.** The dependency between operational and credit losses is achieved through a copula. More specifically, a copula is used to generate two normal standard random variables Y (i.e., credit risk) and Z (i.e., operational risk) via Monte Carlo simulations.
- **Credit loss.** Credit exposures may be divided into categories based on the probabilities of default. The expected loss for each of them is calculated as $PD \times LGD \times A$, where PD is the probability of default, LGD is the loss given default, and A stands for exposure. The expected loss for the portfolio $\mu_{tl,cl}$ is obtained as the sum of the loss of each category. A correlation is also assumed to characterize each category pair. On the basis of these correlations, the portfolio standard deviation ($\sigma_{tl,cl}$) is computed accordingly. Therefore $\mu_{tl,cl}$ and $\sigma_{tl,cl}$ are the only parameters used to simulate the portfolio loss rate (i.e., credit loss divided by the total exposure). A beta distribution (whose moments are characterized by $\mu_{tl,cl}$ and $\sigma_{tl,cl}$) is randomly generated to obtain the credit loss ratio LR_{tl}. Therefore, for each simulation $g = 1, \ldots, G$, the credit loss is

$$CLoss_{tl,h,g} = \sum_{i=1}^{n} A_i \cdot B^{-1}[\Phi(Y_{tl,cl,g})], \qquad (7.6)$$

where $\sum_{i=1}^{n} A_i$ is the credit total portfolio exposure, B^{-1} is the inverse cumulative beta distribution, $\Phi(\cdot)$ is the normal cumulative distribution function, and Y_g is a realization from the above-mentioned standardized normal variable.

- **Operational loss.** According to the BIS (2001), a convenient solution for operational losses relies on a Poisson distribution to generate the number of loss events and a lognormal distribution for their severity. Dimakos and Aas

(2004) use a crude approximation for the overall operational loss based on a lognormal distribution as detailed below:

$$OLoss_{tl,h,g} = e^{(\alpha_{0,tl,ol} + \alpha_{1,tl,ol}.Z_{tl,ol,g})}, \qquad (7.7)$$

whereby $Z_{tl,ol,g}$ is a realization from the above-described standardized normal variable (the subscript ol stands for operational loss). $\alpha_{0,tl,ol}$ and $\alpha_{1,tl,ol}$ are the mean and the standard deviation of the lognormal distribution estimated on the bank-specific historical losses (or indices computed at a regional or sector level).

- **Market loss.** As the last component of the integrated system, the expected return and standard deviation of each trading book instrument are assumed to depend on the size of the credit losses. More specifically, a geometric Brownian motion is used to represent daily fluctuations of market prices. The expected returns and their corresponding standard deviation can be represented as a function of credit losses as follows:

$$\mu_{k,tl,ml,g} = \beta_{0,tl,ml,k} + \beta_{1,tl,ml,k} CLoss_{tl,h,g}, \qquad (7.8)$$

$$\sigma_{k,tl,ml,g} = \gamma_{0,tl,ml,k} + \gamma_{1,tl,ml,k} CLoss_{tl,h,g}, \qquad (7.9)$$

where k stands for a financial instrument within the trading portfolio, while ml indicates market loss. The parameters $\beta_{0,tl,ml,k}$, $\beta_{1,tl,ml,k}$, $\gamma_{0,tl,ml,k}$, and $\gamma_{1,tl,ml,k}$ are fitted on historical data.

- **Integrated loss.** As a final step, the integrated loss is computed by summation of credit, operational, and market losses for each Monte Carlo generation. The value at risk is used to synthesize the overall integrated loss.

Example 7.1 shows how to conduct this approach from a practical point of view.

Example 7.1 Top-Down Top-Level Risk Integration

Let us consider a bank with a $100 billion credit portfolio. The following code shows how to implement the key passages of the top-level mechanism described earlier. The size of the operational losses is assumed to be a proportion of the credit exposure. Additionally, a single asset representing the trading book is used to generate the market risk component.

```
% 1. Normal copula generation of y and z
randn('state', 123456);
rho = 0.4;
nCases = 10e3;
draws = copularnd('Gaussian', rho, nCases);
% 2. Credit loss
A = 100;                              % Exposure
a = 0.4; b = 20;                      % Parameters Beta
CLoss = A * betainv(draws(:,1), a, b);
% 3. Operational loss
alpha_0 = A/4000; alpha_1 = 0.1;      % Paremeters lognormal
OLoss = logninv(draws(:,2), alpha_0, alpha_1);
```

Example 7.1 Top-Down Top-Level Risk Integration—cont'd

```
% 4. Market loss
beta0 = 0.01; beta1 =  0.00003;
gamma0 = 0.01; gamma1 = 0.00002;
mu = beta0 + beta1 * CLoss;
sigma = gamma0 + gamma1 * CLoss;
M1=A/100;
MLoss = M1*exp(mu + sigma.^2 / 2);
% 5. Total loss
TLoss = CLoss + OLoss + MLoss;
VaR=quantile(TLoss,0.99)
```

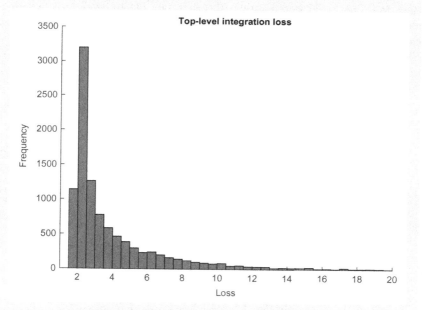

FIG. 7.3 Loss distribution according to the top-level approach ($ billions).

In line with the loss distribution outlined in Fig. 7.3, the value at risk (99% confidence level) represents 16.70% of the credit exposure. The most important contribution comes from the credit risk, which accounts for 14.47%. Then an additional 1.30% is due to the operational risk, and finally the market risk adds 1.01%.

The top-level approach has a series of advantages compared to the basic approach. First of all, losses are simulated on the basis of a dependence structure framework. Additionally, each silo loss imitates the pillar 1 approach. Nonetheless, the key assumptions may not necessarily hold in practice.

The base-level integration process detailed in the next section enhances the overall framework by refining some of the key assumptions of the top-down approach.

7.2.3 Base-Level Integration

One of the key disadvantages of the top-level approach is not to explicitly model economic factors affecting each individual risk source. In contrast, the base-level method aims to transform movements of risk factors into each silo loss via a specific function as shown in Fig. 7.4.

The method is similar to what was discussed for the top-level approach apart from the initial step. Firstly, risk factors are identified and a simulation mechanism is implemented. Then silo losses are computed. Finally, the sum of each risk type is estimated by use of a synthetic indicator (e.g., value at risk). Aas et al. (2007) outline this frame by including ownership and business risk in addition to credit, operational, and market risks. The key steps of the process are as follows:

- **Risk factor simulation.** As a first step, a set of factors influencing the key areas of the integration process are identified. Historical time series are fitted for the market and ownership risks. Then a Monte Carlo architecture is set up to perform conditional simulations. More precisely, the yearly log increments of market risk and ownership risk are transformed into standard normal realizations. Then a new standard normal variable representing the credit risk factor is generated conditionally on the market risk factor. The latter simulation relies on the covariance structure linking market risk and credit risk.

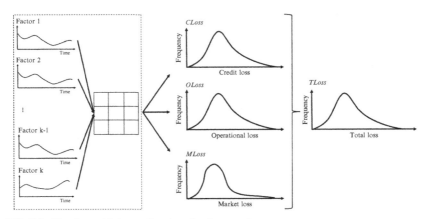

FIG. 7.4 Top-down risk integration: base-level approach.

- **Credit loss.** A similar internal ratings-based (IRB) approach is used for the credit risk. However, three main changes are applied in the IRB framework. A first improvement allows for a diversification among sectors. Secondly, the correlation used for IRB purposes is multiplied by a sector-dependent factor. Finally, an unsystematic risk due to large exposures is taken into account. A sectoral model is based on the aggregation of customers belonging to the same sector s (i.e., $\sum_{i=1}^{n_s} A_{i,s} = A_s$). A simulation process is put in place by use of the above-mentioned credit risk factor (correlated to the market risk factor) and customer-specific normal random variables as detailed below:

$$
CLoss_{bl,h} = \sum_{s=1}^{S} \left[A_s LGD_s \Phi \left(\frac{\Phi^{-1}(PD_s) - \sqrt{\rho_s} \left(\beta_s \xi + \sqrt{1 - \beta_s^2} \vartheta_s \right)}{\sqrt{1 - \rho_s}} \right) \right.
$$
$$
\left. + \sum_{ss=1}^{SS} \mathbb{1}_{s,ss} A_{s,ss} LGD_{s,ss} \right],
$$
(7.10)

where bl indicates base level, and ξ and ϑ_s indicate the credit common factor and the sector-specific random components. Additionally, a concentration contribution is due to $\sum_{ss}^{SS} \mathbb{1}_{s,ss} A_{s,ss} LGD_{s,ss}$, where ss indicates a large exposure (exceeding a given threshold) and $\mathbb{1}_{s,ss}$ assumes the value of 1 when the exposure ss defaults and 0 otherwise.
- **Operational loss.** Similarly to Dimakos and Aas (2004), a lognormal function is used to simulate operational losses.
- **Market loss.** Aas et al. (2007) move from the hypothesis to aggregate portfolio assets into classes linked to a specific factor (e.g., interest rate). A dynamic adjustment of positions during the 1-year holding period is performed as follows. Firstly, they assign a maximum exposure ME_k (defined as the upper utilization limit for the trading portfolio k) and a liquidation period Δ_k for each of these classes. Then the value change of an exposure $\theta_{k,t}$ is based on the log increment of the market risk factor $\gamma_{k,t}$ as follows:

$$
\theta_{k,t} = \begin{cases} 1 - \prod_{r=t}^{t+\Delta_k} e^{\gamma_{k,r}} & \text{if } k \text{ consists of long positions,} \\ |1 - \prod_{r=t}^{t+\Delta_k} e^{\gamma_{k,r}}| & \text{if } k \text{ consists of long and short positions.} \end{cases}
$$
(7.11)

The market portfolio value change for day t is defined as the sum of all asset class variations as detailed below:

$$
MV_t = \sum_{k=1}^{K} ME_k \theta_{k,t}.
$$
(7.12)

The market loss is computed as follows:

$$MLoss_{bl,h} = \max(MV_{t \in h}) \quad \forall MV_{t \in h} \geq 0. \tag{7.13}$$

- **Integrated loss.** The integrated loss is computed as a sum of all risk losses for each factor generation. A statistical measure (e.g., the value at risk) summarizes the integrated loss.

It is worth noting that Aas et al. (2007) also consider ownership and business risk in their base-level integration exercise. On this subject, a specific factor is used for the ownership aiming at computing the value of an insurance company belonging to the bank. The ownership loss is computed as a function of the insurance company. As for the operational risk, a top-down approach is used to aggregate the business risk.

The base-level approach improves the overall integration mechanism. Some key issues on factor interdependencies as well as model parameter estimate remain unresolved. However, it paves the way to a more comprehensive framework based on a bottom-up integration described in the next section.

7.3 BOTTOM-UP ECONOMIC CAPITAL INTEGRATION MODELING

Despite the appealing simplicity and easy-to-implement structure of some top-down approaches, in the last few years bottom-up models have received growing attention. Moreover, the implementation of stress testing procedures helps banks to evolve toward more sophisticated risk integration solutions based on the analysis of their microprocesses.

In what follows, a bottom-up approach is detailed as a continuum with the stress testing framework studied in the previous chapters. Here the focus is on economic capital (i.e., long-run equilibrium). Section 7.4 describes a coherent mechanism to scrutinize the resilience of a bank against short-term liquidity threats.

7.3.1 Economic Capital Integration

A few questions arise when one is dealing with risk integration. More precisely, one should ask: What is the objective function of the process? When does a bank become insolvent? Is there any difference between a long- and a short-run equilibrium? A bank balance sheet representation may help to solve the puzzle.

Let us suppose that at time t the values of a bank's total assets and liabilities are A_t and L_t, respectively. This bank is initially solvent with equity $E_t = A_t - L_t > 0$ (i.e., $A_t > L_t$). The following equation holds to ensure solvency with a probability $(1 - \alpha)$ over a fixed time horizon h (e.g., 1 year):

$$P(E_t + \Delta A - \Delta L + I_h > 0) = 1 - \alpha, \tag{7.14}$$

where ΔA and ΔL are the total asset and liability variations due to macroeconomic instantaneous shocks, and I_h refers to the income for the period under investigation. It is worth noting that the symbol Δ is used throughout the book for two different purposes. On the one hand, it stands for variation (e.g., ΔA). On the other hand, when used as a subscript, it indicates a macroeconomic scenario (e.g., $\mathbf{x}_\Delta$).

It is useful to remark that Kretzschmar et al. (2010) use the simplifying assumption that liabilities are replicated by a portfolio of assets:

$$E(\Delta A) \geq \Delta L - I_h. \tag{7.15}$$

In line with the latter hypothesis and setting $ALoss = -\Delta A$, we can rewrite Eq. (7.14) as follows:

$$P(E_t > ALoss - E(ALoss)) = 1 - \alpha, \tag{7.16}$$

where $ALoss - E(ALoss)$ is the so-called unexpected loss. This argument justifies economic capital assessed at the $(1 - \alpha)$ percentile of the loss distribution described in Chapter 4.

In a more general setting, Bellini (2013) removes the assumption represented in Eq. (7.15) to take into account the contribution of liabilities to bank solvency. In such a context, the asset and liability structure introduced in Chapter 3 is the reference model. A bank with N assets $\mathbf{A}_t = (A_{1,t}, \ldots, A_{N,t})'$ and M liabilities $\mathbf{L}_t = (L_{1,t}, \ldots, L_{M,t})'$ is studied to outline how to integrate market, interest rate, and credit risks. A static balance sheet assumption simplifies the entire integration process. Moving from a similar framework, Section 7.4 focuses on the short-term solvency from a liquidity perspective.

In line with the above economic capital objective, a few alternative risk integration functions may be used for the analysis. In this regard, ΔA, ΔL, and I_h depend on the financial instrument on which a shock is applied as well as the nature of the shock. On the asset side, a distinction holds between the trading book and the banking book. Additionally, ΔA derives from interest rate variations as well as credit defaults. On the liability side, an assumption is made to consider ΔL due to interest rate changes only. Finally, I_h depends on preprovisioning net revenue (i.e., the sum of net interest income, noninterest revenue, and noninterest expenses) and loan impairment charges. A more detailed description is provided in Section 7.2.3. Additionally, in Chapter 8 the choice of the elements to consider becomes crucial to assess the conditions leading a bank to collapse.

The next section describes the process to integrate risks. The focus is on a function encompassing all the components included in Eq. (7.14).

7.3.2 Integration Process

In line with Fig. 7.5, the first step to investigate a bank's resilience from an economic capital perspective is to identify the set of macroeconomic variables affecting its balance sheet and profit and loss. A Monte Carlo simulation process is at the very heart of the entire risk integration method. In more detail, a vector of macroeconomic variables $\mathbf{x}_{\Delta,g}$ is simulated and, for each scenario $g = 1, \ldots, G$, and an integrated loss is computed. The key ingredients of the latter loss are represented in Eq. (7.14).

Asset and liability variations (i.e., ΔA and ΔL) together with the income over a given time horizon I_h are the components of the process. Therefore two main areas of interest are shown in Fig. 7.5. On the one hand, the framework relies on a term structure of interest rates to estimate the present value of both assets and liabilities. On the other hand, asset variations may be caused by defaults. For this reason a transmission mechanism linking macroeconomic variables and credit risk parameters is required. Additionally, a system to generate (correlated) defaults is vital for the entire risk integration architecture.

For each simulation, all instruments are mark-to-model evaluated to assess a fully integrated loss. An integrated loss distribution is then derived, on which a synthetic indicator (e.g., value at risk or expected shortfall) is computed. The key steps of the process are as follows:

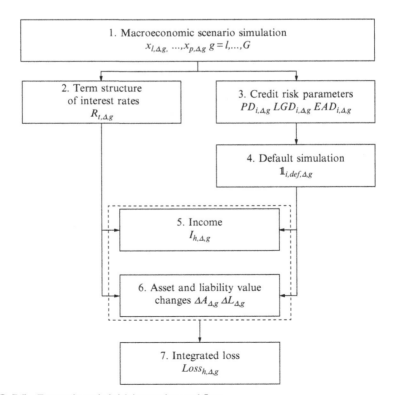

FIG. 7.5 Economic capital risk integration workflow.

1. **Macroeconomic scenario simulation.** Time series analysis is useful not only to fit historical data and project one scenario but also to randomly generate macroeconomic paths $\mathbf{x}_{\Delta,g} = (x_{1,\Delta,g}, \ldots, x_{p,\Delta,g})'$, $g = 1, \ldots, G$ (Bellini, 2016). A Monte Carlo simulation process is crucial for the bottom-up risk integration processes. More specifically, Alessandri and Drehmann (2010) and Drehmann et al. (2010) base their risk integration on vector autoregression-type models. Castren et al. (2010) show how to use a global vector autoregression approach in a multicountry setting. In contrast, one may simulate macroeconomic variables by use of their historical distribution. A multivariate copula or mixture model allows one to capture macroeconomic joint patterns in line with historical occurrences. Following a mixture approach, the density of the macroeconomic variables $\mathbf{x}_t = (x_{1,t}, \ldots, x_{p,t})'$ may be written as follows:

$$g(\mathbf{x}_t) = \sum_{r=1}^{p} \lambda_r \phi_r(x_{r,t}), \tag{7.17}$$

where $(\lambda, \phi) = (\lambda_1, \ldots, \lambda_p, \phi_1, \ldots, \phi_p)$ denotes the vector of parameters, while λ_r are weights and sum to unity. The normal function parametric family, among others, may be a useful starting point to simulate macroeconomic scenarios. All in all, one may alternatively use an econometric model (as detailed in Chapter 2) or a simulation based on a fitted distribution as shown in Eq. (7.17).

2. **Term structure of interest rates.** For each scenario a term structure of interest rates ($R_{t,\Delta,g}$) is required. Firstly, this curve serves as an ingredient to estimate interest revenue and expenses. Then it allows one to estimate asset and liability present value changes due to interest rate shocks. According to Chapter 3, a Vasicek or Cox-Ingersoll-Ross model may be used for fitting purposes. Consequently, the curve is constrained to align with the short-term (r_g^{ST}) and long-term (r_g^{LT}) interest rates generated as part of the macroeconomic scenario $\mathbf{x}_{\Delta,g}$. As an alternative approach, one may refer to Diebold et al. (2006).

3. **Credit risk parameters.** In line with the method described in Chapter 4, credit risk parameters ($PD_{i,\Delta,g}$, $LGD_{i,\Delta,g}$, $A_{i,\Delta,g}$) are shocked at a customer level. They feed both the credit risk loss and the default simulation processes. The assumption of a static balance sheet implies exposures are not to be indexed to macroeconomic variables.

4. **Default simulation.** According to the portfolio modeling discussed in Chapter 4, each customer belongs to a specific sector $s = 1, \ldots, S$. For each macroeconomic scenario a copula is used to generate a random vector $\mathbf{u}_g = (u_{1,g}, \ldots, u_{S,g})'$. Default occurs when $PD_{i,s,\Delta,g} \geq u_{s,g}$ (i.e., $\mathbb{1}_{i,def,\Delta,g}$ assumes the value 1). It is worth noting that copula parameters are required to reflect interdependencies among sector creditworthiness indices. On this subject, Example 7.2 analyzes how to fit a Student T copula on real time series and simulate credit default losses accordingly.

Example 7.2 How to Fit Copula Parameters and Simulate Credit Portfolio Losses

The quarterly Italian default rate time series from 1990 to 2010 may be used as creditworthiness index Ψ_s. A Student T copula is fitted on these time series. Then the five-sector portfolio examined in Chapter 4 is used to simulate a credit loss distribution (see Fig. 7.6) through the following R code:

```
# 1. Load data
library(copula)
t.orig        <- as.data.frame(read.csv("Chap7sector5.csv",
header = TRUE, sep = ";", dec="."))
sector.5      <- as.matrix(read.csv("Chap7sector5head.csv",
header = TRUE, sep = ";", dec="."))
SIMcustomer.5 <- as.data.frame(read.csv("Chap7copula5.csv",
header = TRUE, sep = ";", dec="."))
# 2. Set parameters
set.seed(1234567)
t.d           <-t.orig[,2:6]
mat.cor.td  <- cor(t.d, method="kendall")
cor.td      <- mat.cor.td[upper.tri(mat.cor.td,diag=FALSE)]
n.row.td    <- nrow(t.d)
n.col.td    <- ncol(t.d)
nSimI       <- 10000
# 3. Fit data with  Student-T copula
t.copl      <- tCopula(cor.td,dim=n.col.td, dispstr="un",
df=5,df.fixed=TRUE)
aa          <- rep(0, n.col.td*(n.col.td-1)/2)
u           <- apply(t.d, 2, rank) / (n.row.td + 1)
fit.tau     <- fitCopula(t.copl, u, method="itau")
param.estimate <-attributes(fit.tau)$estimate
fitted.t.copula <- tCopula(param.estimate, dim=n.col.td,
dispstr="un", df=5, df.fixed=TRUE)
# 4. Simulate random generations
r.t.copula      <- rcopula(fitted.t.copula,nSimI)
# 5 Merge simulation with SECTOR
rand.copula     <-cbind(sector.5,t(as.matrix(r.t.copula)))
# 6. Merge database simulation and customer (by row)
mm1             <- as.matrix(SIMcustomer.5$SECTOR)
colnames(mm1)  <- c("SECTOR")
rand.copula.1  <- merge(mm1, rand.copula,
by.x = "SECTOR", by.y = "SECTOR")
rand.copula.pd <-as.matrix(rand.copula.1
[,2: ncol(rand.copula.1)] )
# 7 Simulate defaults
cr.customer     <- 200
```

Example 7.2 How to Fit Copula Parameters and Simulate Credit Portfolio Losses—cont'd

```
sim.default0cr <- matrix(NaN, nrow=cr.customer,ncol=nSimI)
for (i in 1:cr.customer)
for (j in 1:nSimI)
{
  {
  verif <-  SIMcustomer.5[i,4]- rand.copula.pd[i,j]
  if(verif>=0){sim.default0cr[i,j]<-(1-SIMcustomer.5[i,5])}
  else{sim.default0cr[i,j]<- 1}
  }
}
percen       <- 0.999
ptf.value0cr <- t(SIMcustomer.5[1:cr.customer,3])
%*%sim.default0cr
loss0cr      <- sum(SIMcustomer.5[1:cr.customer,3])
-ptf.value0cr
q0cr.999     <- quantile(loss0cr, percen)
UL0cr.999    <- q0cr.999 - mean(loss0cr)
sortloss0cr  <- as.matrix(sort(loss0cr, decreasing=FALSE))
ES0cr.999    <- mean(sortloss0cr[percen*nSimI:nSimI,])
```

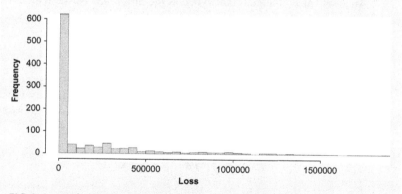

FIG. 7.6 Credit loss distribution based on a Student T simulation ($).

The analysis highlights that $UL_{VaR_{99.9\%}}$ =16.63% of the portfolio value and $UL_{ES_{99.9\%}}$ = 18.26%. A substantial alignment holds with the upper bound loss threshold computed by use of the high correlation hypothesis (i.e., $\rho = 0.5$) described in Chapter 4.

5. **Income.** According to what was described in Chapter 5, the income is represented as follows:

$$I_{h,\Delta,g} = PPNR_{h,\Delta,g} - CLoss_{h,\Delta,g} - Tax_{h,\Delta,g}, \tag{7.18}$$

where $PPNR_{h,\Delta,g} = NII_{h,\Delta,g} + NIR_{h,\Delta,g} - NIE_{h,\Delta,g}$ (where $PPNR$ is the preprovisioning net revenue, NII is the net interest income, NIR is the noninterest revenue, and NIE is the noninterest expenses) and $CLoss_{h,\Delta,g}$ is the credit risk loss. The latter is considered as part of the asset and liability value change described in the next step of the process.

6. **Asset and liability value changes.** Asset and liability full fair value is computed for each scenario by the embedding of both interest rate shocks and credit losses. As detailed in Chapter 3, the present value of a zero coupon instrument is computed as follows:

$$PV = \exp[-R_T(\mathbf{x}_0) \cdot T],$$

where PV is the present value, T is the maturity, and $R_T(\mathbf{x}_0)$ identifies the risk-free interest rate, which is a function of the ongoing (t_0) vector of macroeconomic variables $\mathbf{x}_0$. More generally, $R_t(\mathbf{x}_0)$ indicates the risk-free interest rate in line with $\mathbf{x}_0$, for node t of the term structure.

An asset value change is calculated as the difference between the present value due to shocked macroeconomic variables $\mathbf{x}_\Delta$ and the ongoing $\mathbf{x}_0$. The following equation allows us to compute:

$$\Delta A_{i,\Delta,g} = PVA_{i,\Delta,g} - PVA_i, \tag{7.19}$$

where $PVA_{i,\Delta,g}$ stands for the present value of asset i under the scenario Δ corresponding to simulation g. In contrast, PVA_i indicates the present value of the same asset under $\mathbf{x}_0$. In more detail, the present value of each asset A_i under the current (t_0) economic conditions is computed as follows:

$$PVA_i = \sum_{t=0}^{T_i} PV[A_{i,cf,t}, R_t(\mathbf{x}_0)], \tag{7.20}$$

where PV is estimated by application of Eq. (7.20) on each cash flow $A_{i,cf,t}$ and T_i is the time corresponding to the last cash flow for the ith customer. Interest rate as well as credit risk shocks are taken into account to estimate $PVA_{i,\Delta,g}$ used for risk integration purposes. More precisely, interest rates are shocked because of macroeconomic scenarios. Additionally, default events are generated through $\mathbb{1}_{i,def}$, which is affected by the same macroeconomic variables. For each generation, $g = 1, \ldots, G$, the present value of asset A_i is as follows:

$$PVA_{i,\Delta,g} = \mathbb{1}_{i,def,\Delta,g} \cdot \sum_{t=0}^{T_i} PV[A_{i,cf,t}, R_{t,g}(\mathbf{x}_\Delta)], \tag{7.21}$$

where $\mathbb{1}_{i,def,\Delta,g}$ is a function of the debtor default probability ($PD_{i,\Delta,g}$).

One needs to bear in mind that two main approaches may be followed to jointly take into account interest rate, market, and credit risk. On the one hand, the present value can be computed by use of a risk-free curve, and a default may be simulated as described above. On the other hand, a credit spread can be added to the risk-free interest rate to compute a credit risk-adjusted present value (Grundke, 2009, 2010; Kretzschmar et al., 2010). The latter is strictly conditioned to (market) credit spread availability (e.g., credit default swap).

On the liability side, the following equation holds:

$$\Delta L_{j,g} = PVL_{j,\Delta,g} - PVA_j. \tag{7.22}$$

As per the evaluation of assets, for each liability L_j, the following equation applies:

$$PVL_j = \sum_{t=0}^{T_j} PV[L_{j,cf,t}, R_t(\mathbf{x}_0)]. \tag{7.23}$$

The present value is computed by our assuming PD_{bank} does not change. For each generation, g, the liability's shocked present value is computed according to the following equation:

$$PVL_{j,\Delta,g} = \sum_{t=0}^{T_j} PV[L_{j,cf,t}, R_{t,g}(\mathbf{x}_\Delta)]. \tag{7.24}$$

7. **Integrated loss.** Finally, all the above components are used to compute an integrated loss ($Loss_{h,\Delta,g}$). A quantile or a given measure computed on the integrated loss distribution serves as an economic capital measure. At the end of the above-mentioned process, the integrated total loss function is defined as follows (for each simulation g):

$$Loss_{h,\Delta,g} = -\sum_{i=1}^{N} \left(PVA_{i,\Delta,g} - PVA_i \right) + \sum_{j=1}^{M} \left(PVL_{j,\Delta,g} - PVL_j \right)$$
$$- PNI_{h,\Delta,g} - NIR_{h,\Delta,g} + NIE_{h,\Delta,g} + Tax_{h,\Delta,g} + \Delta NPL_{h,\Delta,g}, \tag{7.25}$$

where $PNI_{h,\Delta,g}$ stands for (simulation g) performing net interest. Performing (nondefaulted) asset interest is compared against all liability interest. $NIR_{h,\Delta,g}$ and $NIE_{h,\Delta,g}$ are noninterest revenue and noninterest expenses, respectively. It is worth mentioning that the usual modeling relies on the performing portfolio. However, the nonperforming portfolio deserves special attention. In particular, in adverse macroeconomic conditions specific provisioning may be deeply affected. Thus $\Delta NPL_{h,\Delta,g}$ is included in Eq. (7.25) with the aim of incorporating the variation of nonperforming losses. The positive sign stands for an increase in nonperforming losses, which contributes to an overall boost of losses.

In the very last step of the process, the integrated economic capital is defined as the value at risk or expected shortfall of the total loss distribution obtained over the predefined holding period h at a given confidence level.

The next section investigates Bank Alpha to grasp the practical implications of the use of an integrated economic capital risk measure.

7.3.3 Bank Alpha's Integrated Economic Capital

It is now interesting to apply the framework described throughout this bottom-up economic capital section to Bank Alpha. The seven-step process summarized in Fig. 7.5 is implemented as listed below:

1. **Macroeconomic scenario simulation.** The global vector autoregression model estimated in Chapter 2 is used to simulate macroeconomic scenarios. The simulation scheme is applied over a multiperiod horizon to generate 1000 coherent random vectors $\mathbf{x}_{\Delta,g}$, $g = 1, \ldots, 1000$. Constraints are imposed with the aim of avoiding unrealistic paths (e.g., deep negative interest rates).

2. **Term structure of interest rates.** A Cox-Ingersoll-Ross model is simulated via the Kalman filter approach described in Chapter 3. For each macroeconomic scenario, short-term and long-term interest rates are considered as constraints to shape the curve.

3. **Credit risk parameters.** Each asset A_i is assigned a probability of default and loss given default at t_0. For each scenario, the shock is on both the probability of default and the loss given default.

4. **Default simulation.** A Student T copula function is estimated by use of a default rate time series on $S = 15$ sectors. Thus 1000 simulated random numbers, $u_{s,g}$, are generated. The default indicator function $\mathbb{1}_{i,def,g}$ is obtained by comparison of $PD_{i,\Delta,g}$ against the copula realization $u_{s,\Delta,g}$. Default occurs when $PD_{i,\Delta,g} > u_{s,\Delta,g}$.

   As detailed in Fig. 7.7, a low negative relationship (i.e., correlation -0.22) links the default rate and interest rates .

5. **Income.** A 3-year period is considered. The performing net interest $(PNI_{h,\Delta,g})$ is computed by use of the characteristics of each financial instrument. Interest rate shocks are applied for each scenario. Without loss of generality, noninterest revenue $(NIR_{h,\Delta,g})$ and noninterest expenses $(NIE_{h,\Delta,g})$ are assumed to be constant. Then a given taxation rate is applied on the profit before tax.

6. **Asset and liability value changes.** Both interest rate and credit risk are taken into account to compute the present value of assets. The credit risk component is captured through the default simulation described above. From the liability point of view, only interest rate changes are considered.

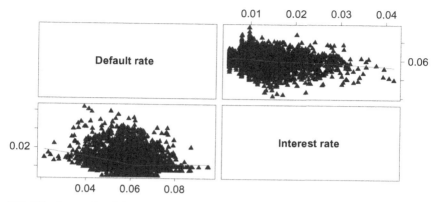

FIG. 7.7 Scatter plot of simulated default rates and interest rates.

Liabilities are separated according to the repricing structure and liquidity features described in Chapter 3.

7. **Integrated loss: value at risk, and expected shortfall computation.** The final step of the integration process is to compute the loss detailed in Eq. (7.25). The value at risk is then computed as a quantile of the loss distribution, while the expected shortfall is the expectation of losses greater than the value at risk. Fig. 7.8 shows the loss distribution. A distinction is made between asset and liability components.

Table 7.1 represents the overall loss value at risk and expected shortfall as well as their subcomponents. More specifically, a \$8.65 billion total value

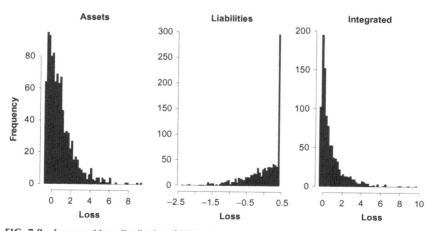

FIG. 7.8 Integrated loss distribution (\$ billions).

TABLE 7.1 Integrated Economic Capital: $EC_{VaR,99.9\%}$ and $EC_{ES,99.9\%}$ ($ Billions)

	Total	Assets		Debits
		Default	Interest	
$EC_{VaR,99.9\%}$	8.65	8.50		0.15
		8.95	−0.45	
$EC_{ES,99.9\%}$	8.90	8.70		0.20
		9.30	−0.60	

at risk can be split between contributions from assets and debits . On this, the former play the major role. This is mainly due to very low interest rates characterizing both sides of Bank Alpha's balance sheet. Additionally, the distinction between default and interest rate contributions within the $8.50 billion asset losses is mainly caused by credit risk. The same reasoning applies to the expected shortfall.

In an attempt to build a parallel between $EC_{VaR,99.9\%}$ and the stress test results, Chapter 6 highlighted that the maximum stress test RWAs due to market and credit risks was $93.63 billion (i.e., $2.76 billion + $90.87 billion = $93.63 billion). The corresponding risk integration RWAs due to market, credit, and interest rate risks (including the profit and loss impact) is $108.13 billion (i.e., $8.65 billion × 12.5 = $108.13 billion). It is worth mentioning that the overall balance sheet structure matters when one is defining a relationship between an integrated loss function against a silos approach. In the case of Bank Alpha, the credit risk is increased because of the high historical correlation among sector creditworthiness indices and a name concentration that are not captured through the IRB formula.

The next section investigates how to integrate risks within a liquidity space.

7.4 BOTTOM-UP LIQUIDITY INTEGRATION MODELING

The Northern Rock case introduced in Chapter 1 highlighted the role of liquidity to ensure banking solvency. As a consequence, Eq. (7.14) needs to be integrated to explicitly take into account liquidity risks. The following equation pinpoints the conditions to ensure solvency with a probability of $(1 - \alpha)$ over the time horizon q (e.g., 3 months):

$$P\left(ALiq_q - LLiq_q + B_q > 0\right) = 1 - \alpha, \tag{7.26}$$

where $ALiq_q$ represents (asset) liquidity inflows over period q, $LLiq_q$ stands for (liability) cash outflows, and B_q corresponds to the liquidity raised through the market. No other liquidity sources are taken into account to simplify the representation. Therefore Eq. (7.26) can be rewritten as follows:

$$P\left[\sum_{t=0}^{q}\left(\sum_{i=1}^{N} \mathbb{1}_{i,liq,t}A_{i,cf,t}(1 - H_{i,t}) - \sum_{j=1}^{m} \mathbb{1}_{j,liq,t}L_{j,cf,t}^{nm}\right.\right.$$
$$\left.\left. - \sum_{j=m+1}^{M} \mathbb{1}_{j,liq,t}L_{j,cf,t}^{m} + \varrho_t B_t\right) > 0\right] = 1 - \alpha, \tag{7.27}$$

where the indicator function $\mathbb{1}_{i,liq,t}$ represents the occurrence of cash inflows $A_{i,cf,t}$, while $\mathbb{1}_{j,liq,t}$ applies to cash outflows $L_{j,cf,t}$. A distinction is made between nonmaturity liabilities ($L_{j,cf,t}^{nm}$) (e.g., current accounts) and instruments with a predefined contractual maturity ($L_{j,cf,t}^{m}$). The first category of liabilities may be withdrawn without any notice. This phenomenon can be related to macroeconomic conditions. Additionally, a liquidity haircut ($H_{i,t} \in [0,1]$) is applied when assets need to be sold to raise cash. This haircut represents the price reduction required by the market for instantaneous liquidity supply. Finally, as detailed in Eq. (7.26), an interbank market buffer (B_t) allows banks to balance cash outflows and inflows. When a crisis occurs, borrowing liquidity from other banks may become difficult. The coefficient $\varrho_t \in [0,1]$ represents the liquidity shrinkage due to macroeconomic adverse conditions (no present value calculation is required because of the short period q).

The next section details how to assess liquidity issues in a risk-integrated framework.

7.4.1 Risk Integration: Liquidity (Short-Term Perspective)

The distribution of liquidity mismatches can be represented as follows:

$$LM_{q,g} = \sum_{t=0}^{q}\left(\sum_{i=1}^{N} \mathbb{1}_{i,liq,t}A_{i,cf,t}(1 - H_{i,\Delta,g,t}) - \sum_{j=1}^{m} \mathbb{1}_{j,liq,\Delta,g,t}L_{j,cf,\Delta,g,t}^{nm}\right.$$
$$\left. - \sum_{j=m+1}^{M} \mathbb{1}_{j,liq,t}L_{j,cf,t}^{m} + \varrho_{\Delta,g,t}B_{\Delta,g,t}\right), \tag{7.28}$$

where $H_{i,\Delta,g,t}$ represents the haircut due to the gth simulation of the macroeconomic vector $\mathbf{x}_\Delta$. The distinction between nonmaturity, $L^{nm}_{j,cf,\Delta,g,t}$, facilities and financial instruments with a cash flow contractual maturity, $L^m_{j,cf,t}$, plays an important role. In fact, liquidity crises may induce customers to withdraw, at least part of, their deposits. For this reason a stochastic component affects $\mathbb{1}_{j,liq,\Delta,g,t}L^{nm}_{j,cf,\Delta,g,t}$ because of macroeconomic conditions. In contrast, no randomness hits liabilities with contractual maturity. Lastly, $\varrho_{\Delta,g,t}$ is the shrinkage associated with that simulation.

Following Amihud (2002), a proxy of market liquidity can be inferred from the stock market. Amihud defines as *ILLIQ* the ratio between the absolute value of a stock's daily return and its daily dollar volume. This ratio shows the price change per daily trading volume for each stock. For a given period DD_{ST} (i.e., month, quarter, and so on) the illiquidity ratio may be computed (for a single stock) as follows:

$$ILLIQ_{ST} = \frac{1}{DD_{ST}} \sum_{dd=1}^{DD_{ST}} \frac{|Re_{ST,dd}|}{Vol_{ST,dd}}, \qquad (7.29)$$

where $Re_{ST,dd}$ is the return on stock ST for day dd and $Vol_{ST,dd}$ is its traded volume (Lu and Glascock, 2010). The average market illiquidity across stocks $ILLIQ_{avg,t}$ (where the subscript t stands for the period DD_{ST} corresponding to a month, quarter, and so on) is calculated as the average of individual stock ratios.

The functional relationship between $ILLIQ_{avg,t}$ and macroeconomic variables is as follows:

$$ILLIQ_{avg,t} = v_0 + v_1 x_{1,t} + \cdots + v_p x_{p,t} + \epsilon_{avg,t}, \qquad (7.30)$$

where ϵ_{avg} is the error component.

With the aim of simulating $ILLIQ_{avg,\Delta}$, the following equation holds:

$$ILLIQ_{avg,\Delta} = \hat{v}_0 + \hat{v}_1 x_{1,\Delta} + \cdots + \hat{v}_p x_{p,\Delta}. \qquad (7.31)$$

A threshold to identify both $H_{i,\Delta,t}$ and $\varrho_{\Delta,g,t}$ is required. When the simulated illiquidity falls below the threshold, no liquidity issue arises. In contrast, when the illiquidity exceeds the boundary, haircut and liquidity shrinkage are generated through uniform random variables, $H_{i,\Delta,g,t} \in [H_{inf}, H_{sup}]$ and $\varrho_{\Delta,g,t} \in [\varrho_{inf}, \varrho_{sup}]$.

$LM_{VaR,(1-\alpha)}$ and $LM_{ES,(1-\alpha)}$ are easily derived once the liquidity mismatching distribution is generated.

Example 7.3 outlines how Eqs. (7.30) and (7.31) are estimated and used to simulate $H_{i,\Delta,t} \in [H_{inf}, H_{sup}]$ and $\varrho_{\Delta,g,t} \in [\varrho_{inf}, \varrho_{sup}]$.

Example 7.3 Haircut $H_{i,\Delta,t} \in [H_{inf}, H_{sup}]$ and Liquidity Shrinkage $\varrho_{\Delta,g,t} \in [\varrho_{inf}, \varrho_{sup}]$ Simulation

The time series of a few of the Italian blue-chip shares are considered over the period from 2003 to 2009 to embrace the recent financial crisis. Quarterly $ILLIQ_{avg,t}$ time series are computed by use of Eq. (7.29). A bank is assumed to have $200 billion of assets split into three main aggregates accounting for $40 billion, $60 billion, and $100 billion. The liquidity buffer B_t is $10 billion. An R code is implemented to show the distribution of assets and the liquidity buffer by following the next steps:

- The illiquidity index is regressed against macroeconomic variables.
- A set of 1000 $ILLIQ_{avg,\Delta}$ is simulated on the basis of a given vector of macroeconomic realizations $\mathbf{x}_\Delta$.
- The illiquidity threshold is set equal to 0.6. This means that if $ILLIQ_{avg,\Delta} > 0.6$, then illiquidity occurs.
- The haircut applied to the vector of assets $\mathbf{A}_t = (40, 60, 100)'$ is in the interval $[0, 0.4]$.
- The liquidity shrinkage applied to the buffer B_t (i.e., $10 billion) stays within the range $[0.4, 0.5]$

```
# 1. Liquidity index regression
database.integration<-as.data.frame(read.csv
("Chap7liquidity.csv",header = TRUE, sep = ";",
dec="."))
lm.illiq<- lm(formula=
ILLIQ ~ INV+CONS, data=illiq.macro.norm)
summary(lm.illiq)
illiq.coef=as.matrix(lm.illiq$coef)
# 2. Liquidity index simulation
sim.macrov<-as.data.frame(read.csv("Chap7simmacro.csv",
header = FALSE, sep = ",", dec="."))
sim.macrov.1<-cbind(matrix(1,nrow=(nrow(sim.macrov)),
ncol=1),sim.macrov[,4:5])
colnames(sim.macrov.1)<-c("INTERCEPT","INV", "CONS")
n.sims=1000
illiq.sim=as.matrix(t(illiq.coef))%*%as.matrix(t(sim.macrov.1))
illiq.sim.norm=(illiq.sim-min(illiq.sim))/(max(illiq.sim)
-min(illiq.sim))
# 3. Illiquidity flag
illiq.sim1<- matrix(0, nrow=1,ncol=n.sims)
threshold=0.6   # Liquidity threshold
```

(Continued)

Example 7.3 Haircut $H_{i,\Delta,t} \in [H_{inf}, H_{sup}]$ and Liquidity Shrinkage $\varrho_{\Delta,g,t} \in [\varrho_{inf}, \varrho_{sup}]$ Simulation—cont'd

```
for (j in 1:n.sims)
{
verif<- illiq.sim.norm[,j]-threshold
 if(verif>0)
 {
 illiq.sim1[,j]<- 1
 }
}
# 4. Haircut applied to assets
sim.inf=0
sim.sup=0.4
illiq.sim.bernoulli<- matrix(0, nrow=3,ncol=n.sims)
for (j in 1:n.sims)
{
verif<- illiq.sim1[,j]-0
        if(verif>0)
        {
        illiq.sim.bernoulli[,j]<- as.matrix(runif(3,
        min=sim.inf, max=sim.sup))
    }
}
asset=as.matrix(c(40,60,100))
asset.sim<- matrix(0, nrow=3,ncol=n.sims)
for (i in 1:3)
for (j in 1:n.sims)
{
    asset.sim[i,j]=asset[i,]*(1-illiq.sim.bernoulli[i,j])
}
asset.sum<-colSums(asset.sim)
# 5. Liquidity shrinking
liq.acc.inf<-0.4
liq.acc.sup<-0.5
liq.access<- 10      # Buffer B_t
market.liq<- matrix( liq.access, nrow=1, ncol=n.sims)
for (j in 1:n.sims)
{
    verif<- illiq.sim1[,j]-0
    if(verif>0)
        {
    market.liq[,j]<- runif(1,min=liq.acc.inf,
    max=liq.acc.sup)* liq.access
    }
}
```

Example 7.3 Haircut $H_{i,\Delta,t} \in [H_{inf}, H_{sup}]$ **and Liquidity Shrinkage** $\varrho_{\Delta,g,t} \in [\varrho_{inf}, \varrho_{sup}]$ **Simulation—cont'd**

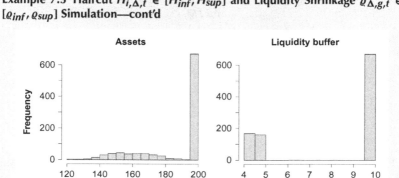

FIG. 7.9 Haircut and shrinkage impact on assets and the liquidity buffer ($ billions).

The haircut causes a reduction in the face value of assets (i.e., $200 billion) when the illiquidity threshold is crossed (Fig. 7.9). At the same time, a shrinkage affects the liquidity buffer (i.e., $10 billion).

In the next section the framework is applied to Bank Alpha.

7.4.2 Bank Alpha's Integrated Liquidity

The illiquidity ratio of Eq. (7.29) is computed by use of the time series of ten blue-chip shares. Starting from these individual ratios, we compute the average market illiquidity $ILLIQ_{avg}$. Thus the regression parameters of Eq. (7.30) are estimated by use of the set of macroeconomic variables detailed in Section 7.3.3. Simulated macroeconomic variables are then used to derive the simulated illiquidity ratio of Eq. (7.31). Therefore $ILLIQ_{\Delta}$ is rescaled so it stays within the range [0,1]. The liquidity threshold is set at 0.6. As a consequence, when a liquidity shock occurs and $ILLIQ_{\Delta}$ exceeds the threshold, both $H_{i,\Delta,t}$ and $\varrho_{\Delta,g,t}$ are randomly generated within the intervals [0,0.4] and [0.4,0.5], respectively. The deposit withdrawn is also simulated when the liquidity threshold above described is triggered. Withdrawal is simulated by use of a uniform variable within the interval [0,0.2]. The latter assumption states that 20% maximum cash repayment is considered with regard to customer deposits.

TABLE 7.2 Liquidity Mismatching: $LM_{VaR,99.9\%}$ and $LM_{ES,99.9\%}$ ($ billions)

	Total
$LM_{VaR,99.9\%}$	−8.90
$LM_{ES,99.9\%}$	−9.15

For each simulated macroeconomic scenario $\mathbf{x}_{\Delta,g}$ the liquidity mismatching $LM_{q,g}$ is computed. $LM_{Var,(1-\alpha)}$ and $LM_{ES,(1-\alpha)}$ described in Table 7.2 are computed by use of a liquidity holding period q of 3 months or less.

Table 7.2 emphasizes that the risk underlying Bank Alpha's asset and liability structure is higher than that shown through the liquidity coverage ratio and the net stable funding ratio. This higher risk couples with the long-term economic capital risk described in the previous section. All in all, integrated risk measures show some additional weaknesses compared with the regulatory measures. Therefore a bank should consider a more comprehensive framework to effectively assess and manage its overall risk profile.

7.5 SUMMARY

This chapter highlighted that bank solvency needs to be assessed both in the long term and in the short term. Two main approaches outlined how to integrate risks from an economic capital standpoint. Following a top-down perspective, a basic integration method relying on correlation or copulas to merge risks estimated on a silo basis was enriched by the introduction of a simulation process to derive the joint distribution of losses (i.e., top-level and base-level integration). Moving to a bottom-up framework, interactions among market, interest rate, and credit risks were studied to point out long term weaknesses. A process similar to a stress test relied on the following key steps: macroeconomic scenario generation, term structure of interest rate estimation, credit risk shock, default simulation, and computation of integrated losses. A similar method applied to integrate liquidity risks. The study of Bank Alpha pointed out the role of risk interdependencies and highlighted bank-specific issues such as sector and name concentration, and liquidity mismatching.

SUGGESTIONS FOR FURTHER READING

Risk integration has only recently attracted the attention of practitioners and researchers. In the top-down approach literature, one may refer to Kuritzkes et al. (2002), Dimakos and Aas (2004), and Aas et al. (2007). In contrast

Grundke (2009, 2010), Kretzschmar et al. (2010), Alessandri and Drehmann (2010), and Bellini (2013) are useful references in the bottom-up risk integration space.

EXERCISES

Exercise 7.1 Let us consider the database Chap7ITmacvar.xlsx referred to the following Italian time series from 1990 to 2010: gross domestic product (GDP), total free on board exports (EXP), total free on board imports (IMP), gross housing investments (INV), internal consumption (CONS), added value for food, beverages, and tobacco (VAL), and unemployment rate (UR). Columns IR1Y, IR3Y, and IR5Y refer to 1-year, 3-year, and 5-year interest rates to be used in Exercise 7.3.

Estimate the model parameters of Eq. (7.17) by use of a normal mixture by use of MATLAB. Randomly generate 1000 replications of this multivariate distribution.

Exercise 7.2 In Exercise 7.1 a multivariate macroeconomic distribution was fitted by use of a mixture of normal random variables. Check the fitting.

Exercise 7.3 Let us consider the database Chap7loss.xlsx referred to a small credit portfolio. Cash flows are assumed to occur within a 3-year period.

- Compute the present value of each exposure by use of the interest rates (IR1Y, IR3Y, IR5Y). Estimate a linear relationship between interest rates and macroeconomic variables of the database Chap7ITmacvar.xlsx. Then, assess the present value changes due to interest rate movements related to macroeconomic scenarios generated in Exercise 7.1.
- Fit the creditworthiness index (IND) against the macroeconomic variables recorded in Chap7ITmacvar.xlsx. Then, estimate each debtor's shocked PD by use of the scenarios generated in Exercise 7.1.
- Apply a normal copula with correlation 0.1 to simulate defaults based on shocked PDs for each simulated scenario.
- Compute the integrated (interest rate and credit risk) portfolio loss.

Solutions are available at www.tizianobellini.com.

REFERENCES

Aas, K., Dimakos, X., Øksendal, A., 2007. Risk capital aggregation. Risk Manage. 9, 82–107.

Alessandri, P., Drehmann, M., 2010. An economic capital model integrating credit and interest rate risk in the banking book. J. Bank. Finance 34 (4), 730–742.

Amihud, Y., 2002. Illiquidity and stock returns: cross-section and time series effects. J. Financ. Mark. 5, 31–56.

Bellini, T., 2013. Integrated bank risk modeling: a bottom-up statistical framework. Eur. J. Oper. Res. 230, 385–398.

Bellini, T., 2016. The forward search interactive outlier detection in cointegrated VAR analysis. Adv. Data Anal. Classif. 10, 351–373.

BIS, 2001. Working Paper on the Regulatory Treatment of Operational Risk, Bank for International Settlements, Basel.

Breuer, T., Jandacka, M., Rheinberger, K., Summer, M., 2010. Does adding up of economic capital for market and credit risk amount to conservative risk assessment?, J. Bank. Finance 34, 703–712.

Castren, O., Dees, S., Zaher, F., 2010. Stress-testing euro area corporate default probabilities using a global macroeconomic model. J. Financ. Stab. 6, 64–74.

Diebold, F., Rudebusch, G., Aruoba, B., 2006. The macroeconomy and the yield curve: a dynamic latent factor approach. J. Econom. 131, 309–338.

Dimakos, X., Aas, K., 2004. Integrated risk modelling. Stat. Model. 4 (4), 265–277.

Drehmann, M., Stringa, M., Sorensen, S., 2010. The integrated impact of credit and interest rate risk on banks: a dynamic framework and stress testing application. J. Bank. Finance 34, 713–729.

Gelman, A., Hill, J., 2007. Data Analysis Using Regression and Multilevel/Hierarchical Models, Cambridge University Press, Cambridge.

Grundke, P., 2009. Importance sampling for integrated market and credit portfolio models. Eur. J. Oper. Res. 194, 206–226.

Grundke, P., 2010. Top-down approaches for integrated risk management: how accurate are they?, Eur. J. Oper. Res. 203, 662–672.

Kretzschmar, G., McNeil, A., Kirchner, A., 2010. Integrated models of capital adequacy—why banks are undercapitalised?. J. Bank. Finance 34 (12), 2838–2850.

Kuritzkes, A., Schuermann, T., Weiner, S., 2002. Risk measurement, risk management and capital adequacy in financial conglomerates, Wharton Financial Institutions Center, University of Pennsylvania.

Lu, R., Glascock, J., 2010. Macroeconomic effects on stock liquidity, University of Cincinnati.

Chapter 8

Reverse Stress Testing

Chapter Outline

An important challenge in designing an effective stress test is to select scenarios sufficiently extreme and plausible to allow one to understand the key weaknesses of a bank.

Reverse stress testing pursues the goal of highlighting circumstances causing the failure of a business. In this regard, both external conditions and bank internal sources of risk need to be considered.

A first step in building a reverse stress test is to specify a suitable objective function to embrace external and internal potential sources of insolvency. The bank's long-term capability and its short-term capability to face its obligations need to be scrutinized.

A bank is exposed to many interdependent risk sources. Therefore, an integrated framework is imperative to represent risk connection in extreme circumstances.

The search for extreme circumstances leading to failure is conducted by use of both qualitative and quantitative statistical models. Specific focus is placed on bank-specific features (e.g., concentration) investigated through what-if analysis and other expert-driven investigations.

Finally, we focus on external macroeconomic conditions under which a bank fails to meet its obligations.

Stress Testing and Risk Integration in Banks. http://dx.doi.org/10.1016/B978-0-12-803590-0.00008-4

KEY ABBREVIATIONS AND SYMBOLS

$LM_{q,g}$	liquidity mismatching over period q for simulation g
$Loss_{h,g}$	integrated loss over holding period h for simulation g: assets and liabilities shocked
$Loss_{h,g}^{CR}$	integrated loss over holding period h for simulation g: only credit shocked
$Loss_{h,g}^{\Delta A}$	integrated loss over holding period h for simulation g: only assets shocked
$\mathbf{x}_t$	vector of macroeconomic variables at time t
$\mathbf{x}^*(\ell)$	macro scenarios causing $Loss$ to exceed a given threshold ℓ
$\mathbf{x}^*(\mathfrak{lm})$	macro scenarios causing LM to be lower than a given threshold $\mathfrak{lm}$
$\mathbf{x}^*(\ell, \mathfrak{lm})$	macro scenarios causing $Loss \geq \ell$ and $LM \leq \mathfrak{lm}$
$\Upsilon^*(\ell)$	internal events causing $Loss$ to exceed a given threshold ℓ
$\Upsilon^*(\mathfrak{lm})$	internal events causing LM to be lower than a given threshold $\mathfrak{lm}$
$\Upsilon^*(\ell, \mathfrak{lm})$	internal events causing $Loss \geq \ell$ and $LM \leq \mathfrak{lm}$
$(\Upsilon, \mathbf{x})^*(\ell, \mathfrak{lm})$	internal events and macro scenarios causing $Loss \geq \ell$ and $LM \leq \mathfrak{lm}$

8.1 INTRODUCTION

Chapters 2–6 concentrated on the key items qualifying a regulatory stress testing exercise. This framework was then extended in Chapter 7 to embrace interdependencies among risks. At this stage, a question arises: Are the stress testing and risk integration frameworks developed so far able to capture the major weaknesses affecting a bank? The high standardization qualifying a stress testing exercise may leave unexplored some areas of risk. The goal of a reverse stress test is to fill this gap.

As detailed in Section 8.2, a reverse stress test aims to structurally identify the most important risks to which a bank is exposed. This removes much of the arbitrariness of the usual stress test based on historical or hypothetical scenarios. As described in Chapter 7, a difference arises between the long- and short-term solvency. When focusing on the long term, our attention is on the quantification of a loss big enough to cause a bank to fail. In contrast, when the focus is driven by a short-term perspective, liquidity mismatching becomes the objective function.

For a bank exposed to multiple risk factors, many different combinations of stress might result in similar losses. Hence, Section 8.3 pinpoints how a bank's specific portfolio composition as well as its asset and liability balance are relevant items to inspect when one is attempting to find core weaknesses. In line with a structural approach, one needs to identify a threshold beyond which a firm collapses. In this regard, one needs to take into account both long- and short-term boundaries. On the one hand, the regulatory capital may be considered as the last line of defense against unexpected losses. On the other hand, liquidity buffers are vital to ensure solvency in the day-by-day banking activity.

In Section 8.4, internal conditions are scrutinized as autonomous or joint source of risk. A what-if analysis pursues the goal of highlighting endogenous circumstances threatening a bank solvency.

Finally, macroeconomic adverse conditions are a typical source of bank weakness. On this subject, Section 8.5 wraps up all the competences acquired throughout the book by exploring macroeconomic conditions that may cause a bank failure.

From a toolkit perspective, a mix of statistical techniques and expert-driven procedures are used throughout this chapter. All instruments developed along the stress testing and risk integration process are used in this final stage of the journey.

8.2 REVERSE STRESS TESTING OBJECTIVE FUNCTION

The definition of a reverse stress test objective is the primary goal of the entire framework. In what follows, a distinction is made between insolvency due to abnormal losses and insolvency due to lack of liquid resources to run the business. Once the scope has been outlined, the focus moves to a function mapping a bank's extreme occurrences and external macroeconomic conditions. The goal is to track a link connecting internal ruinous events and external scenarios.

8.2.1 Reverse Stress Testing: Economic Capital Versus Liquidity Mismatching

In what follows, a series of alternative objective functions describe how to formally identify conditions leading a bank to fail. Two broad threatening event categories are studied. On the one hand, internal features are explored as potential sources of bankruptcy. On the other hand, external economic conditions are investigated while bank operations are considered as given. As a result, a mix of these causes may end in bank insolvency.

- **Internal features (Υ).** The set of asset, liabilities, and other internal characteristics representing a bank's operational system is denoted by Υ. *Loss* and liquidity mismatching (*LM*) are subsets of Υ. Therefore, Υ, *Loss*, and *LM* are jointly considered to identify events causing a bank to fail.
- **Macroeconomic scenarios ($\mathbf{x}$).** A p-dimensional macroeconomic vector representing external conditions is studied when one is assessing potential sources of a bank's insolvency.

Armed with this framework, let us denote the joint distribution function of Υ, *Loss*, *LM*, and $\mathbf{x}$ as $f(\Upsilon, \mathbf{x}, Loss, LM)$. This distribution allows us to conduct a reverse stress test by pursuing two broad objectives. On the one hand, the aim is to investigate internal features capable of causing a bank failure. On the other hand, the research focuses on adverse macroeconomic conditions causing the collapse. Moreover, for each of these two categories a distinction is made between long- and short-term solvency as detailed below:

1. **Internal features** (economic capital): The first way to represent the reverse stress testing objective function is to focus on a what-if scenario affecting a bank's economic capital. In this case, the focus is on internal events causing *Loss* to exceed a given threshold ℓ as detailed below:

$$\Upsilon^*(\ell) = \underset{\Upsilon}{\text{argmax}} \quad f(\Upsilon|\mathbf{x}, Loss \geq \ell, LM). \tag{8.1}$$

In this case, the solution $\Upsilon^*(\ell)$ intercepts the internal events causing a bank's unwillingness to face its obligations. Specific areas of interest include high losses due to single-name default, sector concentration, and so on.

2. **Internal features** (liquidity mismatching): For the liquidity risk the following optimization problem needs to be solved:

$$\Upsilon^*(\mathfrak{lm}) = \underset{\Upsilon}{\text{argmax}} \quad f(\Upsilon|\mathbf{x}, Loss, LM \leq \mathfrak{lm}), \tag{8.2}$$

where the liquidity mismatching threshold $\mathfrak{lm}^1$ may be defined by use of the liquidity distribution described in Chapter 7.

3. **Internal features** (economic capital and liquidity mismatching): When both economic capital and liquidity mismatching are considered, the following equation holds:

$$\Upsilon^*(\ell, \mathfrak{lm}) = \underset{\Upsilon}{\text{argmax}} \quad f(\Upsilon|\mathbf{x}, Loss \geq \ell, LM \leq \mathfrak{lm}). \tag{8.3}$$

4. **Macroeconomic scenarios** (economic capital): A reverse stress test may pursue the goal of identifying macroeconomic scenarios causing *Loss* to exceed a given threshold ℓ. In this case the objective function is written as follows:

$$\mathbf{x}^*(\ell) = \underset{\mathbf{x}}{\text{argmax}} \quad f(\mathbf{x}|\Upsilon, Loss \geq \ell, LM), \tag{8.4}$$

where the solution $\mathbf{x}^*(\ell)$ is the likeliest macroeconomic scenario causing a bank failure due to large economic losses.

5. **Macroeconomic scenarios** (liquidity mismatching): In the short term, liquidity mismatching may trigger a series of events encompassing the potential for a bank failure. The relationship between macroeconomic conditions and internal features is scrutinized as detailed below:

$$(\mathbf{x})^*(\mathfrak{lm}) = \underset{\mathbf{x}}{\text{argmax}} \quad f(\mathbf{x}|\Upsilon, Loss, LM \leq \mathfrak{lm}). \tag{8.5}$$

---

1. Liquidity mismatching is risky when liabilities exceed assets. Hence $\mathfrak{lm}$ has negative value and the threshold is such to identify (high) negative-value liquidity mismatches.

6. **Macroeconomic scenarios** (economic capital and liquidity mismatching): When one is searching for scenarios causing $Loss \geq \ell$ and $LM \leq \mathfrak{lm}$, the following applies:

$$(\mathbf{x})^*(\ell, \mathfrak{lm}) = \underset{\mathbf{x}}{\text{argmax}} \quad f(\mathbf{x}|\Upsilon, Loss \geq \ell, LM \leq \mathfrak{lm}). \qquad (8.6)$$

7. **Internal features and macroeconomic scenarios** (economic capital and liquidity mismatching): In an attempt to merge the above perspectives, the following equation summarizes scenarios and bank-specific events that may cause a bank to become insolvent:

$$(\Upsilon, \mathbf{x})^*(\ell, \mathfrak{lm}) = \underset{(\Upsilon, \mathbf{x})}{\text{argmax}} \quad f(\Upsilon, \mathbf{x}|Loss \geq \ell, LM \leq \mathfrak{lm}). \qquad (8.7)$$

The above reverse stress testing mechanism relies on a system supplying information on integrated measures of risk (both for economic capital and for liquidity) and macroeconomic scenarios. At the same time, all the above optimization functions rely on the definition of a vulnerability threshold as detailed in the next section.

8.3 INTEGRATED RISK MODELING AND VULNERABILITY THRESHOLDS

As detailed in Eqs. (8.1)–(8.7), one needs to use a fully integrated framework to specify the combinations of Υ and $\mathbf{x}$ causing $Loss \geq \ell$ and $LM \leq \mathfrak{lm}$. Additionally, the need to specify ℓ and $\mathfrak{lm}$ thresholds arises. In what follows, risk integration candidate models and vulnerability thresholds are examined.

8.3.1 Long- and Short-Run Risk Integration

The starting point of the analysis is an integrated model capable of capturing interconnections and representing a bank's behavior in the face of adverse conditions (Bellini, 2013). In what follows, three alternative models are explored in line with the methods described in Chapter 7.

- **Economic capital** (assets and liabilities): The first way to represent the *Loss* distribution feeding the objective functions detailed in Section 8.2 is to rely on the fully integrated loss used for economic capital purposes in Chapter 7 as follows:

$$Loss_{h,g} = -\sum_{i=1}^{N} \left(PVA_{i,\Delta,g} - PVA_i \right) + \sum_{j=1}^{M} \left(PVL_{j,\Delta,g} - PVL_j \right)$$
$$- PNI_{h,\Delta,g} - NIR_{h,\Delta,g} + NIE_{h,\Delta,g} + Tax_{h,\Delta,g} + \Delta NPL_{h,\Delta,g}. \qquad (8.8)$$

This function embraces value changes on assets, liabilities, and profit and loss movements due to fluctuations in market, interest rate, and credit risks.

- **Economic capital** (assets only): In line with the hypothesis to be considered as given the liability structure (Kretzschmar et al., 2010), one may shrink the focus on asset losses and profit and loss as detailed below:

$$Loss_{h,g}^{\Delta A} = - \sum_{i=1}^{N} \left(PVA_{i,\Delta,g} - PVA_i \right) - PNI_{h,\Delta,g} - NIR_{h,\Delta,g} \tag{8.9}$$
$$+ NIE_{h,\Delta,g} + Tax_{h,\Delta,g} + \Delta NPL_{h,\Delta,g},$$

where the superscript ΔA stands for value change in assets only.

- **Economic capital** (credits only): An additional way to investigate a commercial bank's vulnerability is to focus only on its credit risk:

$$Loss_{h,g}^{CR} = - \sum_{i=1}^{n} \left(A_{i,\Delta,g} - A_i \right) + Tax_{h,\Delta,g} + \Delta NPL_{h,\Delta,g}, \tag{8.10}$$

where the superscript CR stands for value changes due to credit only. In this equation, debtors $i = 1, \ldots, n$ are taken into account, and $A_{i,\Delta,g}$ is affected by credit risk only. The reason behind this simplified representation is to concentrate on the key portfolio risk sources. Other potential combinations of assets and liabilities should be considered. Nonetheless, the scheme proposed above captures the major challenges for a commercial bank.

- **Liquidity mismatching.** In spite of the long-run economic capital analysis, the analysis may be focused on liquidity. The following equation introduced in Chapter 7 is a useful candidate to perform the study:

$$LM_{q,g} = \sum_{t=0}^{q} \left(\sum_{i=1}^{N} \mathbb{1}_{i,liq,t} A_{i,cf,t} \left(1 - H_{i,\Delta,g,t} \right) - \sum_{j=1}^{m} \mathbb{1}_{j,liq,\Delta,g,t} L_{j,cf,\Delta,g,t}^{nm} \right.$$
$$\left. - \sum_{j=m+1}^{M} \mathbb{1}_{j,liq,t} L_{j,cf,t}^{m} + \varrho_{\Delta,g,t} B_{\Delta,g,t} \right). \tag{8.11}$$

It is worth mentioning that the liquidity coverage ratio and the net stable funding ratio introduced by Basel III constitute a useful corollary to assess a bank's liquidity profile.

The next section investigates the thresholds ℓ and $\mathfrak{lm}$ beyond which a bank becomes insolvent.

8.3.2 Vulnerability Thresholds

According to what was described in Section 8.2, the overall reverse stress testing framework requires a threshold to identify bank failure. Business Case 8.1 helps us understand the role of such a threshold and its practical implications.

Business Case 8.1   Dexia

As a result of the merger of Belgian, Luxemburgian, and French local government finance banks, Dexia became a major player in European local government finance and retail banking in the early 1990s. In the middle of the last financial crisis its total assets were approximately €650 billion.

During the period between 2007 and 2012, the Dexia balance sheet was repeatedly damaged through losses due to the US subprime crisis. Dexia was hit by high refinancing costs coupled with low loan margins. The severe annual losses required a comprehensive public recapitalization through the stakeholder governments of Belgium, Luxemburg, and France in 2008 and 2012. Additionally, in Oct. 2008 and in 2011, Dexia's new unsecured bond issues and interbank deposits had to be enrolled into large public guarantee programs. During that period major structural measures were adopted encompassing sale of operational franchises in a number of countries. The 2012 restructuring plan led to the disposal of the Belgian and Luxemburgian operations: Dexia Bank Belgium and Dexia Bank Internationale. This left the parent company under an orderly resolution plan with €350 billion in residual assets to be managed and disposed of.

From a capital perspective, in 2008, the bank received its first series of capital injections of €6.35 billion. An additional implicit recapitalization measure was €17 billion in asset guarantees provided in 2008. In light of the Greek crisis, at the end of 2012 another capital injection of €5.5 billion was made by national and local governments of the stakeholder countries.

After its first recapitalization in 2008, Dexia was permitted only to fulfill contractual obligations for coupon payments on hybrid capital and subordinated debt, while committing itself to make no early calls. Yet dividend payments and calls or discretionary payments on any of this debt could be made subject to the condition that the core tier 1 ratio would always exceed 10% of the risk-weighted assets.

The risk-weighted asset benchmark of the EU decision was a poor metric for the capital risk of the bank because it entirely ignored the sovereign credit risk that ultimately severely hit the bank. Dexia was chronically undercapitalized, running leverage ratios in the range of 50.

Business Case 8.1 shows that a loss as well as an unbalanced structure undermines a bank solvency. In this regard, a few alternative capital buffer thresholds may be used for reverse stress testing as detailed below:

- **CET1.** This is the tightest trigger one can use for reverse stress testing purposes and, to some extent, it does not completely represent the overall funds on which a bank relies to run its business.
- **Tier 1 capital (inclusive of additional tier 1 capital).** In this case a more extensive definition of own funds is taken into account. Innovative capital instruments sharing common characteristics with core capital are included as an extended line of defense.
- **Total capital.** A broader definition of a capital buffer based on tier 1 capital and tier 2 capital constitutes an additional alternative vulnerability threshold.

All in all, liquidity issues and the overall financing structure need to be taken into account in addition to the thresholds described above. This is emphasized through Business Case 8.2.

Business Case 8.2   Cyprus Popular Bank

In the middle of the first decade of this century, Cyprus Popular Bank was the second largest bank in Cyprus. In 2006 it was consolidated into a group based in Greece that expanded in eastern Europe and Russia in 2007 and 2008. Its total assets approximated €43 billion in 2010. In 2011 the headquarters returned to Cyprus as a consequence of the pressure exerted by the Central Bank of Cyprus.

Through parallel downgrades of Greek and Cypriot securities, in 2011 the bank's assets also became increasingly ineligible for European Central Bank repo operations. The result was a steep increase of the exposure of the Central Bank of Cyprus to Cyprus Popular Bank (i.e., €9.8 billion in Sep. 2012).

Cyprus Popular Bank issued a significant amount of hybrid capital in 2009 and 2010 in the form of contingent convertibles. Hybrid instruments classified as lower tier 2 pursued the goal of compensating investors for the noncall of bonds on their first date and increasing the core tier 1 capital while the government was investing in it.

During the period from 2010 to 2012 the bank suffered deposit withdrawals of more than €7 billion. Only domestic Cypriot deposits from retail customers remained almost unchanged.

After discussions over the extent of creditor participation, the bank was finally dissolved in Mar. 2013 and its good parts were sold to Piraeus Bank (Greece) and Bank of Cyprus.

Cyprus Popular Bank highlights the role of both capital and fund raising. This example together with the examples of Northern Rock, Lehman Brothers, and Dexia highlights the importance of an overall asset liability sustainable structure as well as the need for a strong liquidity regime. This enforces the importance of identifying a vulnerability threshold in terms of liquidity (lm) as listed below:

- **Liquidity mismatch.** A complete integration with the macroeconomic scenario simulation used for economic capital purposes constitutes the major advantage of this risk measure described in Eq. (8.11). Different parameter combinations (e.g., haircut) may be used to strengthen or release liquidity burdens.
- **Regulatory ratios.** Liquidity coverage ratio and net stable funding ratio thresholds may be set equal to the regulatory ones or more restrictive limits can be used.

The definition of the framework through which assess conditions causing a bank to become insolvent is a crucial step of the reverse stress testing process. In Chapter 7, a Monte Carlo simulation process was used to outline the connection between macroeconomic scenarios, a bank's losses, and liquidity mismatching

(Bellini, 2016). For reverse stress testing, a useful starting point is to investigate bank internal sources of risk as detailed in the next section.

8.4 BANK-SPECIFIC DISASTROUS EVENT FACT FINDING

In the previous sections the reverse stress testing problem was described in terms of the objective function and models to assess a bank's behavior under adverse conditions. In what follows, a qualitative analysis is conducted to explore a bank's potential weaknesses. This qualitative analysis is reinvigorated by a more sophisticated quantitative investigation based on a what-if framework. The awareness of organizational deficiencies allows a bank to define a safe way of doing its business. An inspection focusing on the following main areas needs to be conducted: the trading book, banking book, liquidity, and overall financial structure. The following sections detail how to conduct the study.

8.4.1 Trading Book

Some of the most important failures experienced during the 2007–09 crisis pinpoint the influence of joint factors driving insolvency. A qualitative analysis may be conducted in an attempt to uncover trading book shortcomings by focusing on the following key areas:

- **Exposure.** Trading book exposure may be inspected from different angles. A useful starting point is to consider debtor, sector, and risk band (e.g., rating class). More precisely, one of the key questions arising when one is dealing with the trading book is the following: What does unwillingness to repay imply in terms of portfolio single names or group of names? Consider the example of a massive investment in a specific corporation through bonds with different maturities and coupons. What happens to the bank if this corporation defaults? A what-if process that starts from the highest exposures and conjectures the worst scenario may be conducted. As an example, one may select the name with the greatest outstanding balance in the trading book and hypothesize its failure: What are the implications for the bank? Does the failure of the counterpart threaten bank solvency?
- **Financial instrument type.** An equity investment is usually characterized by higher volatility than a bond. Hence one should start from equity exposures when searching for a ruinous path. Nonetheless, bond creditworthiness, market liquidity, and other investment-specific features need to be further inspected to figure out potential solvency issues.
- **Sector.** Contagion or a domino effect is one of the common sources of risk for investors. Therefore one may study the impact of a sector failure on a bank.
- **Counterpart rating.** A poor rating is a symptom of risk. Thus an additional what-if path may be rooted in clustering investments according to their rating class (or probability of default). Assuming the poorest-rated investments collapse allows us to assess the impact of a low-quality portfolio on bank proficiency.

TABLE 8.1 Trading Book Reverse Stress Testing: Illustrative Example of Qualitative Items to be Explored (Each Column is an Independent Silo)

Outstanding Balance (% of Tier 1)	Instrument Type	Sector	Counterpart Rating	Instrument Trading Frequency
[0 – 10)	Bond	Chemical	[AAA;A-]	Intraday
[10 – 20)	Equity	Finance	[BBB+;BB-]	Daily
[20 – 30)	Hybrid	Real estate	[B+;B−]	Weakly
≥ 30)	Derivative	Telecom	[CCC;C]	Monthly
...	...	...	...	...

- **Instrument trading frequency.** Some portfolios are made up of very liquid instruments but this is not always the case. The example of Lehman Brothers highlights the role of opaque investments. Hence assessing the damage caused by low-frequency instruments is a very important exercise in the assessment of potential causes for a bank's collapse.

Table 8.1 exemplifies some of the elements to take into account to make conjectures on events with the potential to cause a bank failure. Illustrative thresholds are also highlighted to pave the way for an effective recognition of portfolio deficiencies. Each item in the columns may be connected with other items in a what-if scheme as detailed in Example 8.1. This example introduces the idea of a heuristic algorithm to search for events that may cause a bank to become insolvent. This kind of research is deeply rooted in expert assessment, and managerial actions will occur according to the bank's sensitivity to the reverse stress testing findings.

Example 8.1 Trading Book Individual (Joint) Default Algorithm

Let us sample from the trading book portfolio and hypothesize the default of an individual debtor or a group of debtors as depicted in Fig. 8.1. The loss given default is assumed to be 100% and a given *Loss* threshold is defined according to Section 8.3.2. The algorithm works as follows:

- Step 1.
  - Select debtor i. The simplifying assumption of one debtor, one financial instrument is followed to avoid unnecessary hurdles.
  - If statement. Verify whether the exposure A_i to debtor i is higher than the bank's bankruptcy threshold. If it is, then the loop stops. Otherwise go to the next step.
- Step 2.
  - Select a couple of debtors i and r.

Example 8.1 Trading Book Individual (Joint) Default Algorithm—cont'd

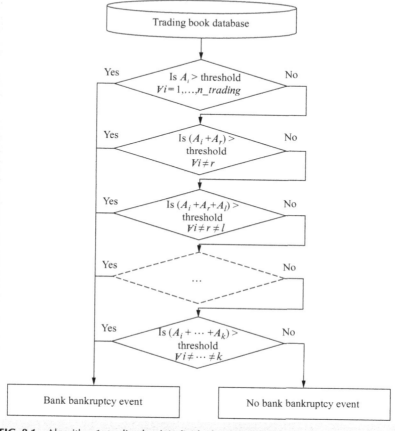

FIG. 8.1 Algorithm 1: trading book individual (joint) default.

- If statement. Verify whether the outstanding balance invested in debtor i and debtor $i \neq r$, $A_i + A_r$, is higher than the bank's bankruptcy threshold. If it is, then the loop stops. Otherwise go to the next step.
- Step 3. Continue the process by considering all sets made up of $k \leq n$ debtors.
- Step 4. All events causing the bank's failure are stored (as detailed in the bank's bankruptcy event box). This set is then scrutinized to assess the relevance of each event. The algorithm does not supply any probability of occurrence; however expert panels may be involved in the process to assess the plausibility of each event.
- Step 5. All plausible events constitute the output of the reverse stress testing process: $\Upsilon^*_{tradingbook}(\ell)$.

A more sophisticated framework may be drawn by the introduction of conditions in terms of rating, sector, exposure, and so on. Example 8.2 uses a poor rating threshold and a sector contagion process.

Example 8.2 Trading Book Poor Rating Sector Contagion Algorithm

Let us focus on the worst layer of the portfolio in terms of rating by introducing a poor rating threshold of BB− (i.e., select customers with a rating of BB− or lower). Additionally assume that a given number (#) of customers in a specific sector (s) default. The maximum number of defaults per sector is set equal to # (i.e., $\max(\sum_{i=1}^{n_s} \mathbb{1}_{(i,s,\leq BB-)}, \#)$). In the case where less than # poor-rating debtors belong to a given sector, all these (poor-rating) debtors are taken into account. Furthermore, hypothesize the loss given default is 60%. A given *Loss* threshold is defined according to Section 8.3.2.

The algorithm described in Fig. 8.2 can be summarized as follows:

- Step 1.
  - Select debtor *i*.
  - If statement. Verify whether the rating for debtor *i* is worse than BB−. If it is, go to the next step. Otherwise select another debtor.

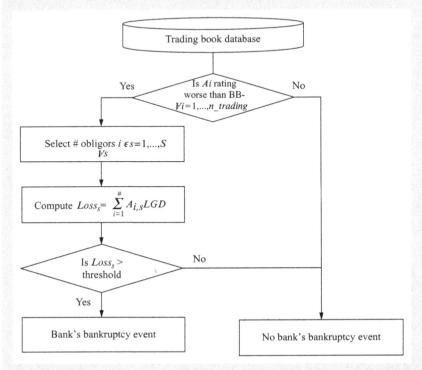

FIG. 8.2 Algorithm 2: trading book poor rating sector contagion. *LGD*, Loss given default.

Example 8.2 Trading Book Poor Rating Sector Contagion Algorithm—cont'd
- Step 2. Select all customers belonging to sector s by aligning with the rule $\max(\sum_{i=1}^{n_s} \mathbb{1}_{(i,s,\leq BB-)}, \#)$.
- Step 3. Compute $Loss_s$ as $\sum_{i=1}^{n_s} A_{i,s} \times LGD$, where LGD is the loss given default. As detailed asbove , a constant loss given default of 60% is assumed.
- Step 4. If $Loss_s \geq threshold$, then a bank failure event is captured.
- Repeat the process for all debtors and all sectors.
- Step 5. All plausible events constitute the output of the reverse stress testing process: $\Upsilon^*_{tradingbook}(\ell)$.

Examples 8.1 and 8.2 can be enriched by the inclusion of different thresholds and modification of the parameters. Moreover, combinations of algorithms may be used to capture real dynamics under stressed conditions.

At step 5 of Examples 8.1 and 8.2, the set of events causing a bank to fail to meet its obligations is specified according to Eq. (8.1). In other words, this set of events is $\Upsilon^*(\ell)$ referred to the trading book (i.e., $\Upsilon^*_{tradingbook}(\ell)$).

The following sections enter into the details of banking book and liquidity algorithms.

8.4.2 Banking Book

An accurate asset review is at the very heart of banking book strategies. Thus an effective diagnosis of name and sector concentration may prevent a bank from being dragged into unexpected tumultuousness. For these reasons a summary of the key risk sources is a useful what-if analysis starting point. For the banking book, exposure class, financial instrument type, sector, and counterpart rating play a key role as detailed in Table 8.2.

TABLE 8.2 Banking Book Reverse Stress Testing: Illustrative Example of Qualitative Items to be Explored (Each Column is an Independent Silo)

Outstanding Balance (% of Tier 1)	Instrument Type	Sector	Counterpart Rating
[0 − 10)	Secured loan	Chemical	[AAA;A−]
[10 − 20)	Unsecured loan	Finance	[BBB+;BB−]
[10 − 20)	Current account	Real estate	[B+;B−]
≥20	Bond	Telecom	[CCC;C]
...	...	...	...

In line with the previous section, a series of combined events may lead to bank insolvency. In what follows, an algorithm that takes into account exposure, sector, and rating is detailed.

Example 8.3 combines a few elements of name and sector concentration. However, the choice of considering as risky all debtors with a rating worse then

---

Example 8.3 Banking Book Name and Sector Concentration

Let us concentrate on both name and sector concentration. A relevant threshold corresponds to 5% of tier 1 capital for the single name and 20% for the sector concentration. With regard to sector concentration, customers with a rating worse than BB− are taken into account. Additionally, assume the loss given default is 60% and a given *Loss* threshold is defined according to Section 8.3.2.

The following steps summarize the algorithm represented in Fig. 8.3:
- Step 1.
  - Select debtor *i*.
  - If statement. Verify whether the outstanding balance is greater than the 5% tier 1 threshold. If it is, then compute the (name concentration) loss by application of a loss given default of 60%. Otherwise go to step 2.
- Step 2. Select all customers in a sector having a rating worse than *BB−*. If the sum of their outstanding balance is greater than 20% of tier 1 capital, then compute the (sector concentration) loss by application of a loss given default of 60%.

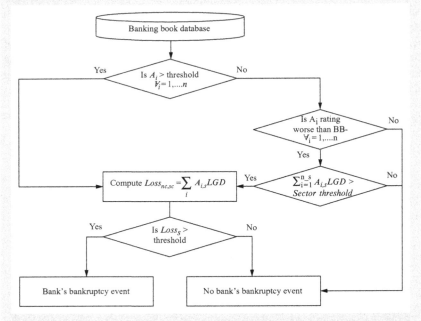

FIG. 8.3 Algorithm 3: banking book name and sector concentration. *LGD*, Loss given default; *nc*, name concentration; *sc*, sector concentration.

Example 8.3 Banking Book Name and Sector Concentration—cont'd

- Step 3. Sum the name and sector concentrations.
- Step 4. If $Loss_s \geq threshold$, then a bank failure event is captured.
- Repeat the process for all debtors and all sectors.
- Step 5. All plausible events constitute the output of the reverse stress testing process: $\Upsilon^*_{bankingbook}(\ell)$.

BB− affects the collection of bank failure events. Setting another rating threshold generates a different set of reverse stress testing events. Additionally, combinations of trading book and banking book events may enrich the spectrum of bankruptcy plausible sources.

Step 5 in Example 8.3 depicts the ruinous event set $\Upsilon^*(\ell)$ referred to the banking book (i.e., $\Upsilon^*_{bankingbook}(\ell)$).

All in all, a broader category of fatal events may be ascribed to structural imbalance between assets and liabilities as detailed in the next section.

8.4.3 Liquidity and Overall Financial Structure

As for the trading book and the banking book, a what-if analysis may help us understand a bank's major risk sources in terms of liquidity. In this regard, one of the key issues experienced by Cyprus Popular Bank and Northern Rock was a deposit run. On this, one needs to bear in mind the distinction between liabilities having a predefined cash flow schedule and deposits without a contractual maturity. In the latter case, a distinction is made between stable and unstable deposits.

Hypothesizing volatility in deposits is the first way of thinking about liquidity as a reason for bank insolvency. Likewise, difficulty in arranging the placement of a bank's own bonds is an important source of financial tension that can eventually cause distress. The same occurs when a short-term misalignment between cash outflows and inflows occurs at the same time as unruliness pervades the interbank liquidity market. In contrast, the asset side may be affected by unwillingness to pay, an increase in credit line usage, and so on.

Bank Alpha is a useful example to investigate liquidity issues causing a bank to become insolvent. Let us focus on Bank Alpha's liquidity ladder. Table 8.3 shows that [0–1M) bucket[2] is marked by a massive negative gap. The following questions arise: What are the consequences of withdrawal of 10% of deposits? Is bank solvency affected by a 20% cash repayment? Under what conditions does liquidity tension cause Bank Alpha to collapse?

On the liability side a distinction is made between instruments having a predefined contractual maturity ($L^m_{j,cf,t}$) and nonmaturity facilities ($L^{nm}_{j,cf,t}$).

---

2. *M* stands for *month*, square brackets include extremes, and parentheses do not include extremes.

TABLE 8.3 Bank Alpha's Liquidity Gap Analysis ($ Billions)

	[0–1M)	[1M–3M)	[3M–6M)	[6M–12M)	[1Y–2Y)	[2Y–5Y)	[5Y–10Y)	≥10Y	Total
Assets	3.00	3.00	4.00	15.00	21.00	34.00	4.00	5.00	89.00
Liabilities	−62.50	−5.00	−2.50	−5.00	−8.00	−8.00			−91.00
Gap	−59.50	−2.00	1.50	10.00	13.00	26.00	4.00	5.00	−2.00

M, Month(s); Y, year(s).

Withdrawals affect the latter, which can also be represented as a function of macroeconomic conditions $L_{j,cf,t}^{nm}(\mathbf{x}_t)$.

Table 8.3 does not supply all the information required to infer Bank Alpha's riskiness from the liquidity perspective. Nonetheless, the [0–1M) bucket liability gap is due to $62.50 billion of deposits. In the case where all deposits are stable, no risk arises. However, a more detailed investigation on deposit stability is needed to check whether the $59.50 billion [0–1M) bucket liability gap is a potential cause of failure. Assuming customers withdraw 5% of their deposits, Bank Alpha's net outflow is $0.13 billion ($3.13 billion liability outflow minus $3.00 billion asset inflows). In the case of a withdrawal of 10% of deposits, the [0–1M) bucket liability gap becomes $3.25 billion. In both cases, Bank Alpha is required to borrow liquidity from other sources. The commonest and quickest way is to turn to the interbank market. In case of a withdrawal of 5% of deposits, $0.13 billion may be raised easily. However, when $3.25 billion is required in a very short time, Bank Alpha may face some difficulties.

In addition, when adverse conditions are taken into account, the entire liquidity profile may be at risk as highlighted through the integrated liquidity mismatching equation in Section 8.3.1.

In particular, one needs to consider interbank market buffer (B_t), the credit liquidity line, and liquidity shrinkage due to macroeconomic adverse conditions (ϱ_t). Let us assume that Bank Alpha can promptly (within 5 working days) activate credit liquidity lines for a maximum of $4.00 billion. A liquidity shock due to deposit withdrawal contagion affecting 5% of the overall nonmaturity liabilities is a substantial threat. In fact, economic conditions as well as reputation implications may induce the interbank market to shrink liquidity. If this contraction achieves 30% of the upper limit, a more complex process is triggered by requiring Bank Alpha to renegotiate its assets. Hence, most liquid assets are due to be sold and a haircut $H_{i,t}(\mathbf{x}_t)$ needs to be applied. Given the nature of the trading book, low haircuts may be applied in the case of asset selling. Thus the hypothesized withdrawal of 5% of deposits does not seem to be harsh enough to cause Bank Alpha to collapse. However, the withdrawal of 10% or 20% of deposits will cast more doubt on the bank's capability to face liquidity issues. Therefore these circumstances can be described as set of events $\Upsilon^*(\text{lm})$.

All in all, the process described above shows the importance of a deep investigation into a bank's liquidity profile. The Basel III requirements may hide fragilities implicit in bank-specific business that can be uncovered through a rigorous reverse stress testing process. Additionally, a combination of economic capital and liquidity mismatching events should be captured in line with Eq. (8.3) (i.e., $\Upsilon^*(\ell, \mathfrak{lm})$).

The next section focuses on macroeconomic scenarios that may compromise bank solvency.

8.5 EXPLORATION OF RUINOUS MACROECONOMIC SCENARIOS

According to the CEBS (2010), *"reverse stress testing consists in identifying a significant negative outcome and then identifying the causes and consequences that could led to such an outcome. In particular, a scenario or combination of scenarios that threaten the viability of the institution's business model is of particular use as a risk management tool in identifying possible combinations of events and risk concentrations within an institution that might not be generally considered in regular stress testing."* Hence the idea of searching for a scenario or combination of scenarios that may cause a bank to collapse is a leading component of an entire risk management framework. As anticipated in the previous sections, a miscellany of external and bank-specific idiosyncrasies may cause insolvency.

The risk integration framework described in Chapter 7 sums up all the risk components described throughout the book. As described in the next section, this framework has the earmark of being the ideal candidate to spot macroeconomic scenarios having a catastrophic impact on a specific bank.

8.5.1 Long- and Short-Run Ruinous Scenarios

The risk integration models detailed in Section 8.3 constitute an ideal framework to identify ruinous scenarios for a given bank. In this regard, the distinction between long- and short-term solvency needs to be taken into account. Additionally, alternative default thresholds can be chosen according to what was described in Section 8.3.2. All in all, a mix of models, thresholds, and inference functions may be used to detect macroeconomic scenarios leading to a bank's collapse. The following components are the necessary ingredients of this process:

(a) **Long run** (economic capital function): The definition of economic capital affects the entire analysis. In line with Section 8.3, the following three alternatives may be pursued:
  - **Fully integrated loss** ($Loss_h$). This is computed by use of the credit risk-adjusted (simulated default) present value of both assets and

liabilities, net interest income, noninterest revenue, noninterest expenses, and nonperforming loss variation.

- **Integrated loss, $Loss_h^{\Delta A}$.** In line with Eq. (8.9), this depends on the asset present value and the liability face value.
- **Credit portfolio loss ($Loss_h^{CR}$).** This loss focuses on credits and is calculated by use of Eq. (8.10).

(b) **Long run** (default correlation hypothesis): According to what was shown in Chapters 4 and 7, default correlation plays a crucial part as a risk source. Hence the above analysis may additionally be enriched by our acknowledging an alternative correlation hypothesis, such as historical estimates, extreme scenarios, and judgmental assessments.

(c) **Long run** (vulnerability threshold): As detailed in Section 8.3.2, the choice of the vulnerability threshold is crucial to identify critical scenarios. In this regard, CET1, tier 1, and total capital are potential candidates, but some other thresholds may be considered in line with the scope of the analysis.

(d) **Short term** (liquidity mismatch): A fully integrated perspective should be followed when one is aiming to consider a comprehensive spectrum of liquidity risks. In contrast, regulatory measures such as the liquidity coverage ratio and the net stable funding ratio may also provide a suitable alternative.

(e) **Short term** (vulnerability threshold): As per the economic capital, even in the case of liquidity analysis, a line needs to be drawn to ascertain when bank insolvency occurs. One may use the percentage of the total asset book value, regulatory thresholds applied to the liquidity coverage ratio, the net stable funding ratio, and so on.

Table 8.4 highlights the mix of ingredients one needs to specify in the long-run solvency framework. The combination (a.1, b.1, c.1) described in Table 8.4 leads to the identification of the scenarios for which $Loss_h$, computed by use of historical default correlation, exceeds the *core capital* threshold. As an

TABLE 8.4 Vulnerability Scenario Setup: Economic Capital, Default Correlation, and Threshold Potential Combinations (Illustrative Example)

	Correlation Hypothesis		Vulnerability Threshold		
	b.1 Historical Correlation	b.2 Hypothetical	c.1 CET1	c.2 Tier 1	c.3 Total Capital
a.1 $Loss_h$	X		X		
a.2 $Loss_h^{\Delta A}$					
a.3 $Loss_h^{CR}$					

TABLE 8.5 Vulnerability Scenario Setup: Liquidity Mismatching Potential Combinations

	Vulnerability Threshold	
	e.1 Total Assets (%)	e.2 Regulatory Threshold
d.1 LM_h	X	
d.2 LCR		X

LCR, Liquidity coverage ratio.

alternative (not depicted in Table 8.4), one may consider the mix (a.2, b.1, c.3). It highlights catastrophic scenarios derived from the combination of $Loss_h^{\Delta A}$, computed by use of historical default correlation, and *total capital* vulnerability threshold.

In the liquidity area, a similar path may be followed. In this case a simpler double-entry representation is drawn as shown in Table 8.5.

One needs to bear in mind that alternative objective functions may be used for reverse stress testing purposes. In particular, according to what was stated in Section 8.2.1, the following options hold:

- **Macroeconomic set** $(\mathbf{x})^*(\ell)$**.** Eq. (8.4) spotlights a scenario causing *Loss* to exceed a given threshold ℓ.
- **Macroeconomic set** $(\mathbf{x})^*(\mathfrak{lm})$**.** Eq. (8.5) focuses on liquidity mismatching lower than a (big) negative threshold $\mathfrak{lm}$.
- **Macroeconomic set** $(\mathbf{x})^*(\ell, \mathfrak{lm})$**.** Eq. (8.6) takes into account both of the above components.

From a graphical perspective, Eqs. (8.4) and (8.5) may be independently drawn as in Fig. 8.4. Nonetheless, economic capital and liquidity mismatching are not two separated and independent components. They represent sides of the same coin. Scenarios causing economic distress may indeed exacerbate liquidity issues or vice versa liquidity adversity may cause an economic crisis. Fig. 8.5 summarizes the idea behind these interactions. In line with Figs. 8.4 and 8.5, one may be interested to infer macroeconomic conditions causing a bank to fail. The next section provides a suitable framework by use of robust contours.

8.5.2 Conditional Mean and Hull Contours

Originating from a market portfolio perspective, Glasserman et al. (2015) investigate how to infer the conditional distribution of trading portfolio losses. Focusing on Eq. (8.4), the first step of this process is to estimate $\mathbb{E}(\mathbf{x}|\Upsilon, Loss \geq \ell, LM)$, the conditional mean of the macroeconomic variables given a loss

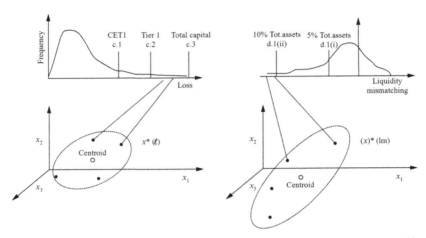

FIG. 8.4 Macroeconomic mapping for independent economic capital and liquidity mismatching outcomes. *Tot.*, Total.

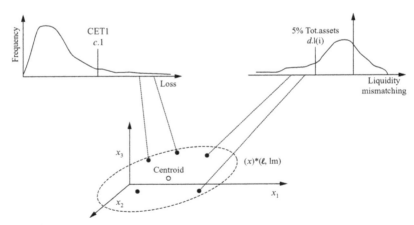

FIG. 8.5 Macroeconomic mapping for joint economic capital and liquidity mismatching outcomes. *Tot.*, Total.

exceeding a given threshold. They rely on observations of past scenarios and their corresponding losses. The combinations characterized by losses exceeding a given threshold ℓ (i.e., $(\mathbf{x}_1, Loss_1), \ldots, (\mathbf{x}_K, Loss_K)$) are taken into account. Then once these observations have been specified, the original problem of estimating a conditional mean reduces to an unconditional mean computation.

They apply an empirical likelihood method that relies on convex combinations of the observations as candidate estimates of the mean:

$$\mathcal{R}(\mathbf{Y}) = \max \left\{ \prod_{k=1}^{K} K w_k : \sum_{k=1}^{K} w_k \mathbf{x}_k = \mathbf{Y} \right\}$$

s.t. (8.12)

$$\sum_{k=1}^{K} w_k = 1,$$

$$w_k \geq 0,$$

where the product inside the braces is the likelihood ratio of the probability vector $(w_1, \ldots, w_K)$ to the uniform distribution $(1/K, \ldots, 1/K)$.

The confidence region for $\mathbf{x}^*$ is defined under specific assumptions on the distribution of observations. In particular, Glasserman et al. (2015) suppose that the observations are i.i.d. with mean μ_0 and the convex hull contains μ_0 with probability approaching 1 as the number of observations increases. A new maximization problem is defined as follows:

$$\operatorname*{argmax}_{w_1, \ldots, w_K} \sum_{k=1}^{K} \log w_k$$

s.t.

$$\sum_{k=1}^{K} w_k = 1,$$ (8.13)

$$\sum_{k=1}^{K} w_k \mathbf{x}_k = \mathbf{Y}.$$

For less regular portfolios and when the number of observations causing *Loss* $\geq \ell$ is small, the above framework hardly applies. However, the idea of searching for a suitable contour of scenarios causing bank failure may be followed. In particular, bivariate box plots have recently been used in robust statistics for the initialization of the *forward search* (Atkinson et al., 2004) and detect atypical units in data envelopment analysis (Bellini, 2012).

A natural nonparametric way of finding a central region in two dimensions is to use convex hull peeling. The output of peeling is a series of nested convex polygons (hulls) that might be fitted through B-spline curves to obtain smooth contours. To find a central part of the data, as described by Zani et al. (1998), a robust bivariate centroid is found on the basis of the observations inside the inner region defined by the fitted splines. In this way both the efficiency property of the arithmetic mean and the natural trimming offered by the hulls are used.

The plots calculated from B-splines are overelaborate to find a central part of the data. Therefore it can be useful to use a simpler method in which ellipses with a robust centroid are fitted to the data as in Riani and Zani (1997). The robust centroid of the ellipse is found as the componentwise median of the two variables in the scatter plot. The shape of the contours is based on a covariance matrix in which the univariate medians are used but which is otherwise calculated in the usual way. The combination of centroid and covariance estimates gives a Mahalanobis distance for each observation and a family of ellipses that need to be scaled.

Example 8.4 shows how to use the MATLAB function to obtain the box plot from ellipses shown in Fig. 8.6.

Example 8.4 Box Plot From Ellipses

In what follows, the MATLAB code uses the Forward Search Data Analysis toolbox to obtain the box plot from ellipses (function *unibiv*). The analysis is based on 20 random generations from a three-dimensional multivariate normal variable.

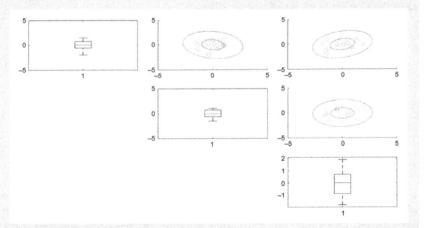

FIG. 8.6 Box plot from ellipses.

```
% Random normal sample generation
randn('state', 123456);
n=20;
p=3;
Y=randn(n,p);
% Show robust confidence ellipses
out=unibiv(Y,'plots',1,'textlab',1);
```

The region embraced by the ellipses represents the area of interest for reverse stress testing purposes. The next section focuses on Bank Alpha to show how the process may be implemented in practice.

8.5.3 Bank Alpha's Ruinous Scenario Analysis

In line with the ruinous scenario analysis objectives, one needs to compute the long-run loss of Eqs. (8.8)–(8.10) and the liquidity mismatching detailed in Eq. (8.11). From the economic capital perspective, the focus is on Bank Alpha's overall asset and liability integrated loss (i.e., Eq. 8.8).

Macroeconomic scenarios are generated by use of the framework described in Chapter 7. Then, three alternative vulnerability thresholds are used: CET1, tier 1, and total capital. The percentage of core tier 1 overrun is 3.90%. The probability to hit the other two limits is 1.50% and 0.50%, respectively. Moving to the contour analysis detailed in Section 8.5.2, first one needs identify scenarios exceeding the vulnerability thresholds. Then, Table 8.6 shows the corresponding centroids. They are substantially in line with the 2007–09 crisis. The size of ellipsoids reduces as far as the vulnerability threshold grows (i.e., CET1, tier 1, and total capital).

From a liquidity point of view, two options are explored. The first relies on $H_{i,\Delta,t} \in [0.0, 0.4]$, $\rho_{\Delta,t} \in [0.4, 0.5]$, and $\gamma_{\Delta,t} \in [0.2, 0.6]$. The second one is based on $H_{i,\Delta,t} \in [0.0, 0.4]$, $\rho_{\Delta,t} \in [0.0, 0.5]$, and $\gamma_{\Delta,t} \in [0.0, 0.6]$. Given a 4% total asset liquidity mismatch threshold, there is a 5.10% probability to overcome the limit when one relies on option one. The probability reduces to 1.50% when option two is considered. From Table 8.6 one may conclude that scenarios causing Bank Alpha to collapse are similar to the recent crisis.

8.6 SUMMARY

Reverse stress testing was introduced as a managerial and regulatory framework through which to explore weakness sources. A few alternative objective functions were investigated by our pointing out the difference between internal features and macroeconomic conditions causing a bank to become insolvent. A bunch of integrated risk models were inspected in an attempt to uncover all major bank weaknesses. Additionally, vulnerability thresholds were investigated from long-term (economic capital) and short-term (liquidity) perspectives.

What-if analyses were proposed as a qualitative tool for inspection of the trading book, banking book, and overall liquidity vulnerabilities. Examples of heuristic algorithms were introduced to clarify how to conduct the (quantitative) computational analysis. Finally, macroeconomic scenarios were investigated as a source of potential bank failure. Interactions between external and internal conditions were studied. The relevance of a fully integrated framework was highlighted as a challenge for a renewed risk management process.

TABLE 8.6 Bank Alpha's contour analysis. Macroeconomic variable centroids for scenarios exceeding vulnerability thresholds.

Vulnerability threshold	US Real GDP Growth Rate (%)	US Unemployment Rate (%)	US Inflation Rate (%)	US Short-Term Interest Rate (%)	US Long-Term Interest Rate (%)	China and DA Real GDP Growth Rate (%)	China and DA Inflation Rate (%)	China and DA USD ER (a)	Euro Area Real GDP Growth Rate (%)	Euro Area Inflation Rate (%)	Euro Area USD ER	Japan Real GDP Growth Rate (%)	Japan Inflation Rate (%)	Japan USD ER	UK Real GDP Growth Rate (%)	UK Inflation Rate (%)	UK USD ER
Economic capital																	
CET1	−3.80	7.00	3.70	0.12	1.10	2.54	5.95	72.80	−5.05	2.50	1.04	−7.55	−1.45	86.40	−3.04	−0.28	1.10
Tier 1	−3.70	7.05	3.80	0.10	1.30	2.80	5.80	74.90	−4.98	4.10	1.03	−7.02	−1.20	85.70	−3.40	−0.10	1.15
Common equity	−3.73	6.98	3.85	0.15	1.12	2.92	6.02	82.60	−4.20	3.50	1.10	−6.89	−1.55	87.12	−2.95	0.10	1.30
Liquidity mismatching																	
Liquidity opt 1	−4.02	6.90	3.60	0.12	1.13	1.90	5.90	75.60	−5.00	3.75	1.15	−7.20	−1.20	85.45	−3.02	−0.15	1.17
Liquidity opt 2	−3.90	6.70	3.75	0.13	1.15	1.85	6.05	78.20	−4.90	3.80	1.10	−7.01	−1.10	86.05	−2.97	−0.10	1.13

(a) The exchange rate (ER) is an index.

SUGGESTIONS FOR FURTHER READING

Reverse stress testing has recently been introduced by regulators and few contributions come from the banking literature. Grundke and Pliszka (2015) and Glasserman et al. (2015) are the most up-to-date references.

EXERCISES

Exercise 8.1 Let us consider an algorithm to figure out bank liquidity weaknesses by focusing on deposit funding. Consider the illustrative example described in Section 8.4.3 to draw a diagram representing all steps of the process with the aim of specifying $\Upsilon^*(\mathrm{lm})$.

Exercise 8.2 In line with the integration between the long-term (economic capital) and the short-term (liquidity mismatching) perspectives, draw a diagram to identify $\Upsilon^*(\ell, \mathrm{lm})$ by use of Example 8.3 and Exercise 8.1 as references for the analysis.

REFERENCES

Atkinson, A.C., Riani, M., Cerioli, A., 2004. Exploring Multivariate Data with the Forward Search. Springer, New York.

Bellini, T., 2012. The forward search outlier detection in data envelopment analysis. Eur. J. Oper. Res. 216, 200–207.

Bellini, T., 2013. Integrated bank risk modeling: a bottom-up statistical framework. Eur. J. Oper. Res. 230, 385–398.

Bellini, T., 2016. The forward search interactive outlier detection in cointegrated VAR analysis. Adv. Data Anal. Classif. 10, 351–373.

CEBS, 2010. CEBS guidelines on stress testing (gl32). CEBS Guidelines.

Glasserman, P., Kang, C., Kang, W., 2015. Stress scenario selection by empirical likelihood. Quant. Finance 15, 25–41.

Grundke, P., Pliszka, K., 2015. A macroeconomic reverse stress testing. Discussion Paper No. 30/2015. Deutsche Bundesbank.

Kretzschmar, G., McNeil, A., Kirchner, A., 2010. Integrated models of capital adequacy—why banks are undercapitalised. J. Bank. Finance 34 (12), 2838–2850.

Riani, M., Zani, S., 1997. An iterative method for the detection of multivariate outliers. Metron 55, 101–117.

Zani, S., Riani, M., Corbellini, A., 1998. Robust bivariate boxplots and multiple outlier detection. Comput. Stat. Data Anal. 24, 257–270.

Index

Note: Page numbers followed by *b* indicate boxes, *f* indicate figures, *t* indicate tables and *np* indicate footnotes.

Printed in the United States
By Bookmasters